Solutions Manual

Perspectives on Structure and Mechanism in Organic Chemistry

Felix A. Carroll
Davidson College

Brooks/Cole Publishing Company
I(T)P® An International Thomson Publishing Company

Pacific Grove • Albany • Belmont • Bonn • Boston • Cincinnati • Detroit • Johannesburg • London
Madrid • Melbourne • Mexico City • New York • Paris • Singapore • Tokyo • Toronto • Washington

Sponsoring Editors: Elizabeth Barelli Rammel, Beth Wilbur
Editorial Associate: Nancy Conti
Production Coordinator: Dorothy Bell

Cover Design: Roy R. Neuhaus
Cover Illustrator: Kenneth Eward/BioGrafx
Printing and Binding: Patterson Printing

For more information, contact:

BROOKS/COLE PUBLISHING COMPANY
511 Forest Lodge Rd.
Pacific Grove, CA 93950
USA

International Thomson Publishing Europe
Berkshire House 168-173
High Holborn
London WC1V 7AA
England

Thomas Nelson Australia
102 Dodds Street
South Melbourne, 3205
Victoria, Australia

Nelson Canada
1120 Birchmount Road
Scarborough, Ontario
Canada M1K 5G4

International Thomson Editores
Seneca 53
Col. Polanco
11560 México, D.F., México

International Thomson Publishing GmbH
Königswinterer Strasse 418
53227 Bonn
Germany

International Thomson Publishing Asia
221 Henderson Road
#05-10 Henderson Building
Singapore 0315

International Thomson Publishing Japan
Hirakawacho Kyowa Building, 3F
2-2-1 Hirakawacho
Chiyoda-ku, Tokyo 102
Japan

Printed in the United States of America

10 9 8 7 6 5 4 3 2 1

ISBN 0-534-34096-2

Acknowledgments

This solutions manual provides literature references and detailed solutions for most of the problems in *Perspectives on Structure and Mechanism in Organic Chemistry*. Some problems do not have literature references. Other problems are designed to stimulate thought and debate, so there is no single correct solution for them.

I am most grateful to David L. Dillon of the University of Wyoming for carefully checking all of the solutions and for diplomatically offering suggestions for revision of many of them. I also wish to express my appreciation for the support and guidance of the staff of Brooks/Cole publishing company, particularly Harvey Pantzis (Executive Editor), Elizabeth Rammel (Ancillaries Editor), Beth Wilbur (Editorial Associate) and Jamie Sue Brooks (Senior Production Editor).

With the exception of some graphs produced with a spreadsheet program and one figure reproduced from a journal, the graphics in this manual were created with ChemWindow (SoftShell International, Ltd., Grand Junction, CO). Figures were saved as WPG files and imported into WordPerfect for DOS. This approach resulted in graphics that differ slightly from those that would have been produced by working entirely within a Windows environment, but because of my hardware and my familiarity with WordPerfect for DOS, this was the most convenient method for me. The stereochemical drawings follow the conventions discussed on pages 83 - 84 of the text.

Felix A. Carroll

Davidson College

Contents

Chapter 1

1. An answer to this question should be stated in terms of macroscopic phenomena, and a historical exposition provides a rationale for the basis of contemporary chemistry. Reference to any of several monographs on the history of chemistry can be used to summarize the ideas and observations that led to contemporary chemistry theory.[1,2,3]

2. See, for example, *Chem. Eng. News* **1986** (Sept. 1), pp. 4-5.
   a. Scanning tunneling microscopy.
   b. The eye sees a computer screen or plotter output. The human eye does not see atoms.
   c. The experiment is based on the premise that matter is made of atoms.
   d. The atoms appear to be bright spots, approximately round in shape.

3. a. The alternative geometries and their elimination on the basis of number of isomers are as follows:
      i. square planar - There would be two isomers of CH_3Cl, one "cis", in which the Cl-C-Cl bond angle is 90°, and one "trans", in which the Cl-C-Cl bond angle is 180°.
      ii. square pyramid - there would be two isomers of CH_3Cl again.

   b. In all answers, a substituent is presumed to replace a hydrogen atom in the parent structure of the candidate structure for benzene.[4]
      i. If benzene had the structure we now call fulvene, there should be three different derivatives with the formula C_6H_5Cl.
      ii. If benzene had the structure we now call Dewar benzene, there would be two and only two isomers with the formula C_6H_5Cl.
      iii. If benzene had the structure we now call benzvalene, there would be three possible isomers with the formula C_6H_5Cl.
      iv. If benzene had the structure we now call prismane, there would be only one isomer with the formula C_6H_5Cl, but there would be four isomers with the formula $C_6H_4Cl_2$ (two of them existing as a pair of enantiomers).

---

[1] Asimov, I. *A Short History of Chemistry*; Anchor Books: Garden City, NY, 1965.

[2] Ihde, A. J. *The Development of Modern Chemistry*; Harper & Row: New York, 1964.

[3] See, for example, Butterfield, H. *The Origins of Modern Science, 1300-1800*, Revised Edition; The Free Press: New York, 1965.

[4] For a discussion of the number of isomers of benzene, see Reinecke, M. G. *J. Chem. Educ.* **1992**, *69*, 859 and references therein.

v. If benzene had the structure we now call [3]radialene, there would be one and only one isomer with the formula C_6H_5Cl, but there would be four possible isomers with the formula $C_6H_4Cl_2$.

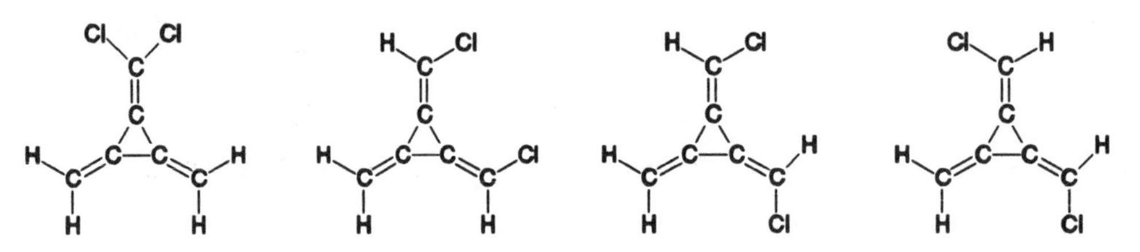

vi. There are also acyclic structures with the formula C_6H_6, such as 2,4-hexadiyne, and they may be analyzed similarly. For example, if benzene were 2,4-hexadiyne, then there would be one and only one C_6H_5Cl, but there would be only two structures with the formula $C_6H_4Cl_2$.

c. One can never know that something that has not been tested is like something else to which it seems similar. However, it seems unproductive to dwell on the possibility until there is an experimental result that could be rationalized on the basis of a structure for chloromethane that is different from the tetrahedral structure of methane. The spectroscopic results for chloromethane are consistent with a tetrahedral geometry.

4. The data and equations are given in Bondi, J. *J. Phys. Chem.* **1964**, *68*, 441. For *n*-pentane, the volume is given by

$$V_W = 2 \times 13.67 + 3 \times 10.23 = 58.03 \text{ cm}^3 \text{ mol}^{-1}$$

and the area is given by

$$A_W = 3 \times 1.35 + 2 \times 2.12 = 8.29 \times 10^9 \text{ cm}^2/\text{mol}^{-1}$$

These results agree with those given by the general formulas for *n*-alkanes:

$$V_W = 6.88 + 10.23N_c = 6.88 + 10.23 \times 5 = 58.03 \text{ cm}^3 \text{ mol}^{-1}$$

$$A_W = 1.54 + 1.35N_C = 1.54 + 1.35 \times 5 = 8.29 \times 10^9 \text{ cm}^2 \text{ mol}^{-1}$$

For isopentane,

$$V_W = 3 \times 13.67 + 10.23 + 6.78 = 58.02 \text{ cm}^3 \text{ mol}^{-1}$$

$$A_W = 3 \times 2.12 + 1.35 + 0.57 = 8.28 \times 10^9 \text{ cm}^2 \text{ mol}^{-1}$$

For neopentane,

$$V_W = 4 \times 13.67 + 3.33 = 58.01 \text{ cm}^3 \text{ mol}^{-1}$$

$$A_W = 4 \times 2.12 + 0 = 8.48 \text{ x } 10^9 \text{ cm}^2 \text{ mol}^{-1}$$

Note that these equations do not consider effects of crowding. A semi-empirical calculation suggests that molecular area decreases along the series *n*-pentane, isopentane, neopentane.

5. Kiyobayashi, T.; Nagano, Y.; Sakiyama, M.; Yamamoto, K.; Cheng, P.-C.; Scott, L. T. *J. Am. Chem. Soc.* **1995**, *117*, 3270.

   81.81 + 29.01 = 110.82 kcal/mol.

6. Turner, R. B.; Goebel, P.; Mallon, B. J.; Doering, W. v. E.; Coburn, Jr., J. F.; Pomerantz, M. *J. Am. Chem. Soc.* **1968**, *90*, 4315. Also see Hautala, R. R.; King, R. B.; Kutal, C. in *Solar Energy: Chemical Conversion and Storage*; Hautala, R. R.; King, R. B.; Kutal, C., eds.; Humana Press: Clifton, NJ, 1979; p. 333.

   The difference in heats of hydrogenation indicates that quadricyclane is less stable than norbornadiene by 24 kcal/mol, so this is the potential energy storage density for the photochemical reaction.

7. Pilcher, G.; Parchment, O. G.; Hillier, I. H.; Heatley, F.; Fletcher, D.; Ribeiro da Silva, M. A. V.; Ferrão, M. L. C. C. H.; Monte M. J. S.; Jiye, F. *J. Phys. Chem.* **1993**, *97*, 243.

$$C_8H_{12}O_{2\,(s)} \rightarrow C_8H_{12}O_{2\,(g)} \qquad \Delta H_s = 23.71 \text{ kcal/mol}$$
$$8\ CO_{2\,(g)} + 6\ H_2O_{(l)} \rightarrow C_8H_{12}O_{2\,(s)} + 1O_{2\,(g)} \qquad -\Delta H_c = 1042.90 \text{ kcal/mol}$$
$$8\ C_{(graphite)} + 8\ O_{2\,(g)} \rightarrow 8\ CO_{2\,(g)} \qquad \Delta H_f = 8(-94.05) = -752.4 \text{ kcal/mol}$$
$$6\ H_{2\,(g)} + 3\ O_{2\,(g)} \rightarrow 6\ H_2O_{(l)} \qquad \Delta H_f = 6(-68.32) = -409.92 \text{ kcal/mol}$$

$$8\ C_{(graphite)} + 6\ H_{2\,(g)} + O_{2\,(g)} \rightarrow C_8H_{12}O_{2\,(g)} \qquad \Delta H_f = -95.71 \text{ kcal/mol}$$

8. See Davis, H. E.; Allinger, N. L.; Rogers, D. W. *J. Org. Chem.* **1985**, *50*, 3601.

$$\Delta H_f\ (\text{phenylethyne}) = \Delta H_f\ (\text{phenylethane}) - \Delta H_r\ (\text{phenylethyne})$$
$$= 7.15 - (-66.12) = 73.27 \text{ kcal/mol}$$

9. Wiberg, K. B.; Hao, S. *J. Org. Chem.* **1991**, *56*, 5108.

$$\Delta H_{r\ (\text{cis-3-methyl-2-pentene})} = \Delta H_{r\ (\text{2-ethyl-1-butene})} - \Delta\Delta H_f = -10.66 - (-1.65) = -9.01 \text{ kcal/mol}$$

10. Fang, W.; Rogers, D. W. *J. Org. Chem.* **1992**, *57*, 2294.

$$\textit{cis}\text{-1,3,5-hexadiene} + 3\ H_2 \rightarrow \textit{n}\text{-hexane} \qquad \Delta H = -81.0 \text{ kcal/mol}$$
$$\textit{n}\text{-hexane} \rightarrow \text{1,5-hexadiene} + 2\ H_2 \qquad \Delta H = +60.3 \text{ kcal/mol}$$

$$\textit{cis}\text{-1,3,5-hexadiene} + H_2 \rightarrow \text{1,5-hexadiene} \qquad \Delta H_r = -20.7 \text{ kcal/mol}$$

$$\textit{trans}\text{-1,3,5-hexadiene} + 3\ H_2 \rightarrow \textit{n}\text{-hexane} \qquad \Delta H = -80.0 \text{ kcal/mol}$$
$$\textit{n}\text{-hexane} \rightarrow \text{1,5-hexadiene} + 2\ H_2 \qquad \Delta H = +60.3 \text{ kcal/mol}$$

$$\textit{trans}\text{-1,3,5-hexadiene} + H_2 \rightarrow \text{1,5-hexadiene} \qquad \Delta H_r = -19.7 \text{ kcal/mol}$$

11. See Smyth, C. P. in *Physical Methods of Chemistry*, Vol. 1, Part 4; Weissberger, A.; Rossiter, B. W., eds.; Wiley-Interscience: New York, 1972; pp. 397-429. The gas phase dipole moments for CH_3-F, -Cl, -Br, and -I are 1.81, 1.87, 1.80 and 1.64 D, respectively. Using the bond length data in Table 1.1 and rewriting equation 1.18 leads to the following partial charges on F, Cl, Br, and I, respectively: -0.27, -0.22, -0.19, -0.16. The dipole moments do not show a monotonic trend along the series because a dipole moment is a **product** of two terms. In series of methyl halides, one term (the partial charge) goes down and the other term (bond length) goes up. The product of these two terms is a maximum at the second member of the series. Note that the assumption that only the carbon and halogen atoms are charged is an oversimplification. An Extended Hückel calculation indicates that the three methyl hydrogen atoms bear some charge also.

12. Because Pauling electronegativities are computed from the properties of atoms in molecules, they generally cannot be computed for the inert gases. However, krypton and xenon fluorides are known, and electronegativities of krypton and xenon have been reported by Meek, T. L., *J. Chem. Educ.* **1995**, *72*, 17.

13. Owen, N. L.; Sheppard, N. *Trans. Faraday Soc.* **1964**, *60*, 634 reported that the syn conformer is more stable than the anti conformer by 1.15 kcal/mol. An answer to this question can be based on the discussion by Walters, E. A. *J. Chem. Educ.* **1966**, *43*, 134 (who cited this experimental result). Essentially, one treats the rabbit ear lone pairs on oxygen as substituents, just as the C-H bonds on the methyl group of propene were treated as substituents in the discussion in the text. A conformation in which the rabbit ear lone pairs are not eclipsed with the bent bond curves of the double bond is lower in energy than is a conformation in which the bent bonds are eclipsed with the unpaired electrons. Therefore, the lower energy conformation is the one in which the R attached to the oxygen in the ether is on the same side of the molecule and in the same plane as is the carbon-carbon double bond.

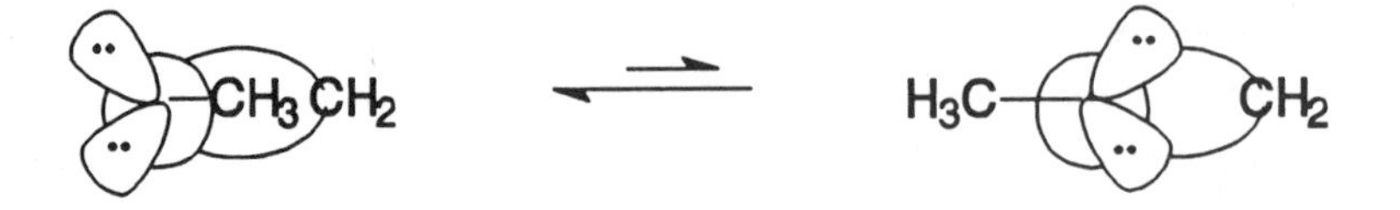

A word of caution is in order here, however. The drawing of nonbonded electrons on oxygen (which is central to the analysis by either model) represents an assumption that may not be warranted. For example, the nonbonded electrons on oxygen on methanol give patterns of electron density in higher calculations that do not show discrete rabbit ear lone pairs. (See the discussion of the lone pairs of water by Liberles, A. *J. Chem. Educ.* **1977**, *54*, 479.) Also, Bond, D.; Schleyer, P. v. R. *J. Org. Chem.* **1990**, *55*, 1003, concluded that ..."It is the molecular dipole that is found to be the primary factor in determining the syn preference... Also contributing to the stability of the preferred syn conformer is an increased conjugation of the oxygen perpendicular lone pair with the double bond in the syn conformation."

14. Using equation 1.41,

$$1 + \lambda_C^2 \cos\theta_{CC} = 0$$

leads to a value of 2.62 for λ_C^2. Therefore the hybridization of carbon orbitals used for carbon-carbon bonds is $sp^{2.62}$. Now using the relation

$$2\left[\frac{1}{1+2.62}\right] + 2\left[\frac{1}{1+\lambda_H^2}\right] = 1$$

leads to a value of 3.47 for the carbon orbitals used for the carbon-hydrogen bonds.

15. Mastryukov, V. S.; Schaefer III, H. F.; Boggs, J. E. *Acc. Chem. Res.* **1994**, *27*, 242. Also see the discussion in Gilardi, R.; Maggini, M.; Eaton, P. E. *J. Am. Chem. Soc.* **1988**, *110*, 7232.

a. As the bond angle increases, the C-C bond length decreases. Conversely, as the bond angle decreases, the C-C bond length increases.

b. The larger α, the greater the contribution of *p* character to the orbital of C2 used for the C2-C3 bond. This means greater *s* character in the orbital of C2 used for the C1-C2 bond, which results in a shorter C1-C2 bond. The same result can be rationalized using the VSEPR approach. As the angle α increases, there is less repulsion in the electrons comprising the C1-C2 bond with the electrons in the C2-C3 bond. This allows the electrons in the C1-C2 bond to move closer to C2, thus decreasing the bond length.

16. The calculation for CH_3Br can be done in the same way as the example of CH_3Cl given in the text. Using the values reproduced in Table 1.1,

$$\lambda_H^2 = -\frac{1}{\cos\theta_{HH}} = 2.77$$

so the carbon orbital used for carbon-hydrogen bonding is $sp^{2.77}$ hybridized. Then

$$3\left[\frac{1}{1+2.77}\right] + \frac{1}{1+\lambda_{Br}^2} = 1$$

so $\lambda_{Br}^2 = 3.90$, and the carbon orbital used for carbon-bromine bonding is $sp^{3.90}$ hybridized. According to the VSEPR model, the more electronegative bromine pulls electron density from the carbon atom, which in turn pulls the electrons in the carbon-hydrogen bonds closer to the carbon nucleus. As a result, the carbon-hydrogen electron pairs repel each other more strongly than they do the pair of electrons in the carbon-bromine bond, so the H-C-H bond angle opens up, and the H-C-Br bond angles close from the expected 109.5°.

17. According to the bent bond formulation, the electrons in the bent bonds are pulled in toward the other olefinic carbon atom, so the electrons in these bonds repel the electrons in the carbon-hydrogen bonds less than they would in propane. Therefore the H-C-H bond angle opens to a larger value.

The electrons in formaldehyde should be pulled even more strongly away from the carbon atom than is the case in ethene. Therefore, the repulsion of electrons in either of these bonds with the electrons in either carbon-hydrogen bond is even less than that in ethene, so the H-C-H bond angle in formaldehyde should be even greater than that of ethene.

18. Robinson, E. A.; Gillespie, R. J. *J. Chem. Educ.* **1980**, *57*, 329 (appendix, p. 333), using an H-C-H angle of 116.2° for ethene, report $sp^{2.26}$ or 30.6% *s* character for the carbon-hydrogen bond. Using 117° for the H-C-H angle[5] leads to $sp^{2.20}$, or 31.2% *s* character. For formaldehyde, using an H-C-H angle of 125.8°[6] similarly leads to λ^2_{HH} of 1.71, which means 36.9% *s* character for the carbon orbital used for carbon-hydrogen bonding.

19. a. The formula is given by Newton, M. D.; Schulman, J. M.; Manus, M. M. *J. Am. Chem. Soc.* **1974**, *96*, 17. Set equation 1.46 = J = 5.7*(%*s*) - 18 Hz. Then $500/(1+\lambda^2)$ = 5.7 × (%*s*)-18. Then let %*s* = $100/(1+\lambda^2)$ and solve for λ^2. It turns out to be just under 3. Thus the equation is approximately correct for orbitals that are roughly sp^3-hybridized, but it is not exact for other orbitals.

b. The equation is

$$r_{C\text{-}H} = 1.1597-(4.17 \times 10^{-4})(500)/(1 + \lambda^2)$$

so

$$r_{C\text{-}H} = 1.1597 - 0.209/(1 + \lambda^2)$$

This equation is equivalent to

$$r_{C\text{-}H} = 1.1597 - 2.09 \times 10^{-3} \times \rho_{C\text{-}H}$$

where $\rho_{C\text{-}H}$ is percent *s* character, which is defined as $100/(1 + \lambda^2)$. This is the form of the equation given by Muller, N.; Pritchard, D. E. *J. Chem. Phys.* **1959**, *31*, 1471.

---

[5](a) *Tables of Interatomic Distances and Configuration in Molecules and Ions*;," Bowen, H. J. M.; Donohue, J.; Jenkin, D. G.; Kennard, O.; Wheatley, P. J.; Whiffen, D. H., comps.; Special Publication No. 11, Chemical Society (London): Burlington House, W.1, London, 1958. (b) Supplement, 1965, p. M 78s.

[6]Reference 5(b), p. M 109.

20. Here are calculations based on literature values for H-C-H bond angles and assuming that all molecules have planar carbon skeletons. (That is necessarily true only for cyclopropane.) Note that the values calculated depend on the choice of literature values for the bond angles.

	cyclopropane[7]	cyclobutane[8]	cyclopentane[9]
<H-C-H	118°	114°	109.5°

(Using the formula $1 + \lambda_i^2 \cos\theta = 0$,)

	cyclopropane	cyclobutane	cyclopentane
λ_i^2 =	2.13	2.459	2.996
Fract. *s* in C-H	0.319	0.289	0.25
Fract. *p* in C-H	0.681	0.711	0.75

Each carbon has 2 C-H bonds and 2 C-C bonds. Therefore for a C-C bond of cyclopropane, the fractional *s* character is 0.5 * (1-2*(0.319)) = 0.181. Similarly,

	cyclopropane	cyclobutane	cyclopentane
Fract. *s* in C-C	.181	.211	.25
Fract. *p* in C-C	.819	.789	.75
λ_j^2 =	4.525	3.74	3.00
C-C-C interorbital angle:	102.77°	105.5°	109.47°

If the molecules are flat, then cyclopropane has (102.77 - 60)/2 = 21.4° of angle strain at each carbon. Similarly, cyclobutane has 7.75° of strain, and cyclopentane has no strain.[10] As will be discussed in Chapter 3, cyclobutane and cyclopentane are not flat. The large fraction of *p* character in the cyclopropane carbon-carbon sigma bonds suggests that they might react (at least to some extent) like π bonds, which is partially true. Note that the inter*orbital* bond angle of cyclopropane is 102.77°, whereas the inter*nuclear* bond angle is required to be 60°. Thus the cyclopropane bonds are bent or banana bonds.[11] The acidity values can be correlated with *s* character by combining equations 1.46 and 1.48 to show a relationship between kinetic acidity and *s* character, and the results shown in Table 1.10 are consistent with such a relationship. By using the VSEPR concept, the very bent carbon-carbon bonds of cyclopropane (and to a lesser extent, cyclobutane) allow the electrons in the carbon-hydrogen bonds to be pulled closer to the carbon nucleus. That not only increases the H-C-H bond angle, but it also stabilizes a carbanion resulting from proton removal, so the acidity of a compound with more bent bonds is greater than that of a compound with less bent bonds.

---

[7]Reference 5(b), p. M98s.

[8]Reference 5(a), p. M 168.

[9]Reference 5(a), p. M 185.

[10]This result for cyclopentane is based on the H-C-H bond angle reported in the literature. If the five carbon atoms of cyclopentane form a perfect pentagon, then the C-C-C bond angles are all 108°, so there is a slight amount of angle strain.

[11]Note also that cyclopropane has been described in terms of Walsh orbitals, which are based on *p* orbitals.

21. See Kass S. R.; Chou, P. K. *J. Am. Chem. Soc.* **1988,** *110*, 7899.

    $1 + \lambda^2 = 500/202$, so $\lambda^2 = 1.475$. Therefore, the percent *s* character is $100/2.475 = 40.4\%$. This is less than the 50% *s* character in acetylene C-H bonds, so acetylene should be more acidic.

22. There is no literature reference for this problem, nor is there a single right answer. One answer is that the concept of hybridization provides a useful conceptual model for understanding the bonding of carbon compounds without the need for carrying out molecular orbital calculations in which hybridized orbitals are not assumed. Therefore hybridized orbitals are useful, but only if it is remembered that they are only a mental convenience and not a physical attribute of atoms.

23. There is no literature refernce for this problem. If hybridization does not exist, then a quantification of hybridization is only a convenient artifice. Artifice it may be, but convenient it is. Even though λ cannot be observed directly, it is a useful concept because it provides a more satisfying conceptual basis for correlating coupling constants, acidities and bond angles with each other than would a purely empirical correlation of any two of these observables.

24. The answer to this question depends on the orientation of the answerer. Organic chemists use pictorial representations because they work with structures that are often larger and more complex than those that can be described in purely mathematical terms — at least those that can be described in such a way that the organic chemist is able to visualize the results. Although organic chemistry may become more mathematical as the role of computation becomes even more important, it is likely that the results of a mathematical analysis will continue to be presented in a largely pictorial "graphical interface" for the organic chemist.

25. Coulson has stated the basic paradox of chemistry. We live in a "macroscopic" world, but we explain that world in terms of unseen particles and unseen forces. To the chemist, atoms, bonds, and molecules are real and can be demonstrated. However, Coulson's reminder that these concepts are intangible reinforces the view that chemistry is based on models that are subject to revision if better models become available.

Chapter 2.

1. a) (*Z*)
   b) *trans*
   c) (1*Z*, s-trans, 3*E*)
   d) (*Z*) or *syn*
   e) *exo*
   f) *rac-trans*
   g) (1*R**,3*R**,5*R**)
   h) ß

2. Structures **A** and **B** are positional isomers because the methoxy group is on C2 in **A** but is on C1 in **B**. Structures **A** and **C** are enantiomers. Structures **A** and **D** are functional group isomers because **A** is a methyl ether, while **D** is an alcohol. Structures **A** and **E** are diastereomers.

3. a) (*R*)-2-chloro-3-oxopropanoic acid
   b) (*R*)-3-bromo-3-phenylpropene
   c) (3*R*,4*S*)-3-bromo-4-chloro-3-methylhexane
   d) (2*S*,3*S*)-2-methoxy-2,3-butanediol
   e) (2*R*,3*S*)-4-chloro-2,3-dihydroxybutanal
   f) (*S*)-1,3-dichloro-1,2-propadiene
   g) (1*S*,2*S*)-1,2-dichlorocyclohexane
   h) (*R*)-2-bromo-2′-chlorobiphenyl
   i) (1*R*,2*R*)-1,2-dimethylcyclopropane
   j) (*R*)-1,2-cyclononadiene (Moore, W. R.; Anderson, H. W.; Clark, S. D.; Ozretich, T. M. *J. Am. Chem. Soc.* **1971**, *93*, 4932.)
   k) (*R*)-1-(bromomethylidene)-4-methylcyclohexane (Gerlach, H. *Helv. Chim. Acta* **1966**, *49*, 1291.)
   l) (*R*)-6,6′-dimethyl-2,2′-diphenic acid (Newman, P.; Rutkin, P.; Mislow, K. *J. Am. Chem. Soc.* **1958**, *80*, 465.)

4. Because the helix makes a clockwise turn as it proceeds away from the observer, it is a *P* helix, which corresponds to the (*S*) designation.

5. Seebach, D.; Lapierre, J.-M.; Skobridis K.; Greiveldinger, G. *Angew. Chem., Int. Ed. Engl.* **1994**, *33*, 440.
   4096. The number of possible stereoisomers is 2^n, where n is the number of chiral centers. There are twelve chiral centers, each marked with a star in the figure.

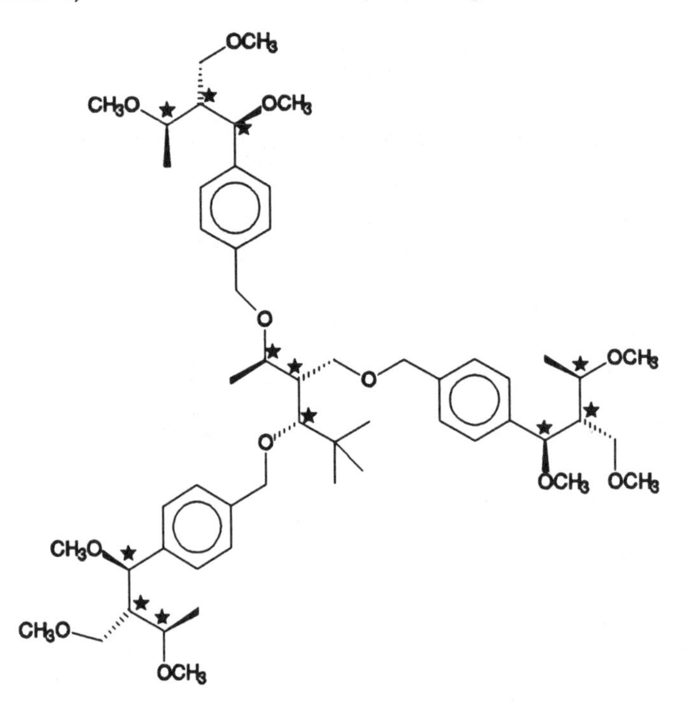

6. c) Reynolds, K. A.; Fox, K. M.; Yuan, Z.; Lam, Y. *J. Am. Chem. Soc.* **1991**, *113*, 4339.
   As drawn on the page, the left face of a) is the *si* face. Similarly, the left face of b) is the *re* face and the front face of c) is the *si* face.

7. d) Reynolds, K. A.; Fox, K. M.; Yuan, Z.; Lam, Y. *J. Am. Chem. Soc.* **1991**, *113*, 4339.
   The protons in a), c) and d) are heterotopic, stereoheterotopic and enantiotopic, while those in b) are heterotopic, stereoheterotopic and diastereotopic. As the structures are drawn, in a) the circled hydrogen on the left is pro-(*S*). In b) the circled hydrogen on the left is pro-(*R*). In c) the circled hydrogen on top is pro-(*S*). In d) the circled hydrogen in front is pro-(*S*).

8. a) D_3, chiral: Eaton, P. E.; Leipzig, B. *J. Org. Chem.* **1978**, *43*, 2483; b) C_2 chiral; and c) D_4, chiral: Halterman, R. L.; Jan, S.-T. *J. Org. Chem.* **1991**, *56*, 5253. d) C_2, chiral: See the discussion by Deprés, J.-P.; Morat, C. *J. Chem. Educ.* **1992**, *69*, A232.

9. a. Cywin, C. L.; Webster, F. X.; Kallmerten, J. *J. Org. Chem.* **1991**, *56*, 2953.

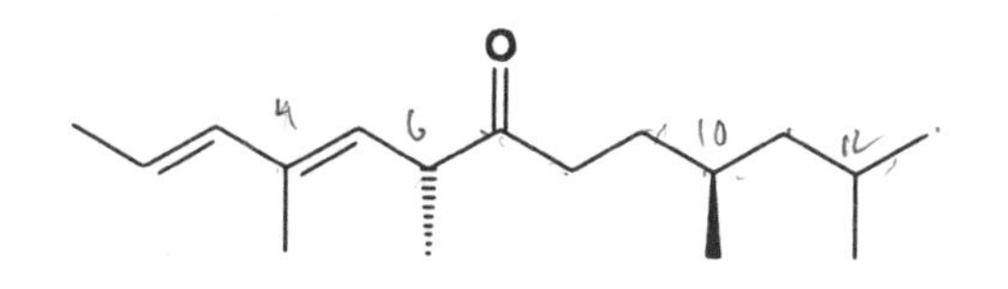

b. Ingold, K. U. *Aldrichimica Acta* **1989**, *22*, 69.

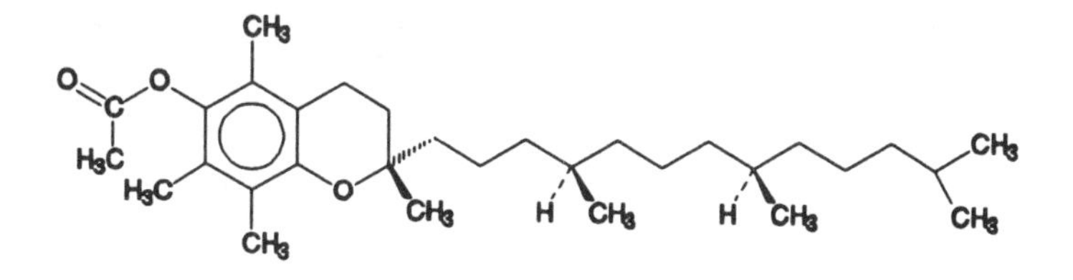

c. Rychnovsky, S. D.; Griesgraber, G.; Zeller, S.; Skalitzky, D. J. *J. Org. Chem.* **1991**, *56*, 5161.

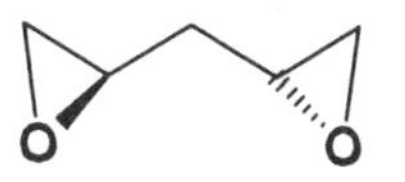

d. Cianciosi, S. J.; Ragunathan, N.; Freedman, T. B.; Nafie, L. A.; Baldwin, J. E. *J. Am. Chem. Soc.* **1990**, *112*, 8204.

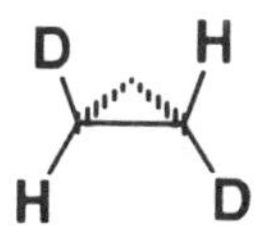

e. This structure was illustrated in an Eastman Fine Chemicals advertisement in *J. Org. Chem.* **1992**, *57* (20), on a page preceding the table of contents.

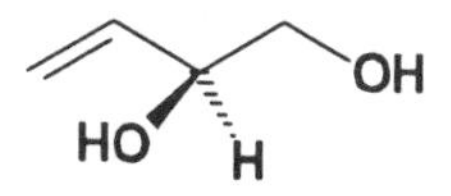

f. This structure was illustrated in an Eastman Fine Chemicals advertisement in *J. Org. Chem.* **1992**, *57* (20), on the page preceding the table of contents.

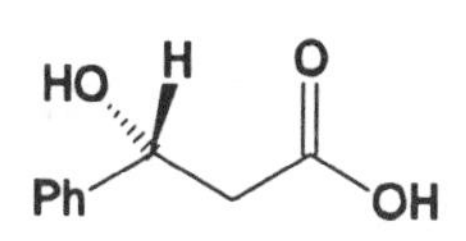

g. Bharucha, K. N.; Marsh, R. M.; Minto, R. E.; Bergman, R. G. *J. Am. Chem. Soc.* **1992,** *114*, 3120.

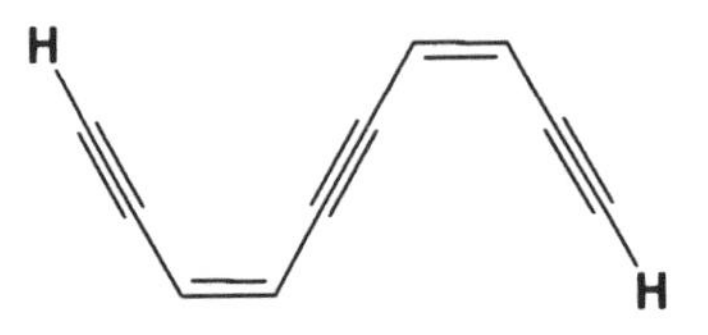

h. Naoshima, Y.; Munakata, Y.; Yoshida, S.; Funai, A. *J. Chem. Soc. Perkin Trans. 1* **1991**, 549.

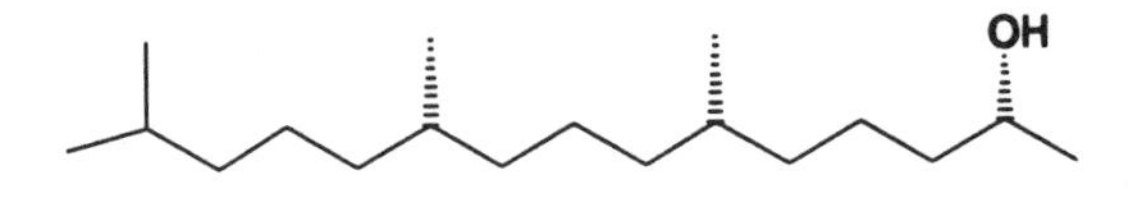

i. Walborsky, H. M.; Goedken, V. L.; Gawronski, J. K. *J. Org. Chem.* **1992**, *57*, 410.

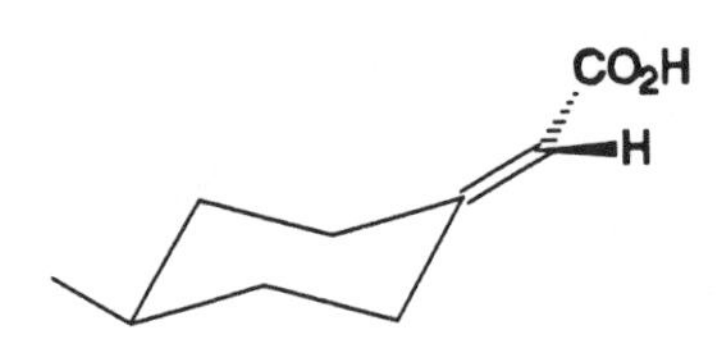

j. Rawson, D.; Meyers, A. I. *J. Chem. Soc., Chem. Commun.* **1992**, 494.

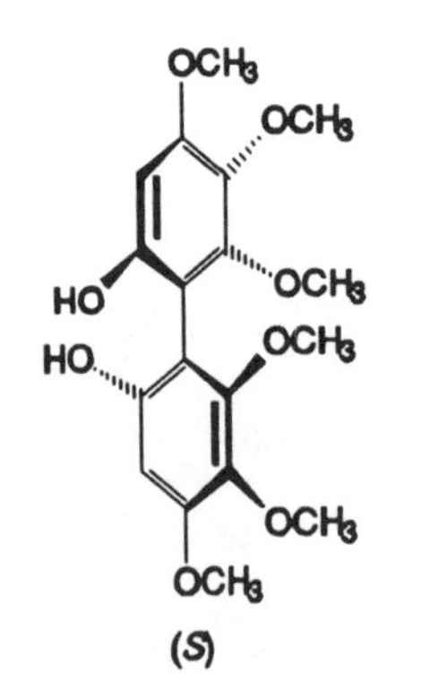

k. Kitching, W.; Lewis, J. A.; Perkins, M. V.; Drew, R.; Moore, C. J.; Schurig, V.; König, W. A.; Francke, W. *J. Org. Chem.* **1989**, *54*, 3893.

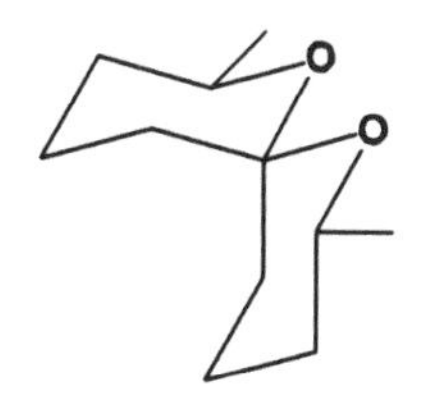

l. Freedman, T. B.; Cianciosi, S. J.; Ragunathan, N.; Baldwin, J. E.; Nafie, L. A. *J. Am. Chem. Soc.* **1991**, *113*, 8298.

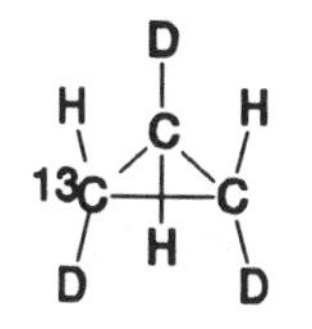

m. Rao, A. V. R.; Gurjar, M. K.; Bose, D. S.; Devi, R. R. *J. Org. Chem.* **1991**, *56*, 1320.

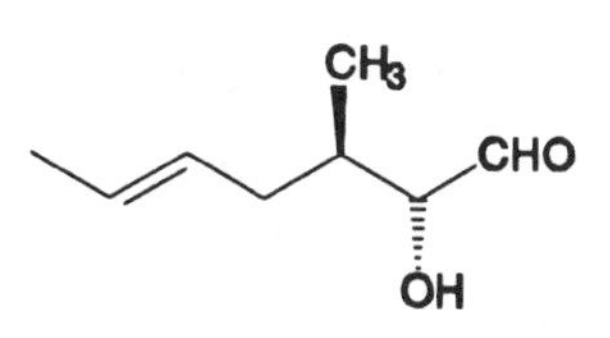

n. Liu, C.; Coward, J. K. *J. Org. Chem.* **1991**, *56*, 2262.

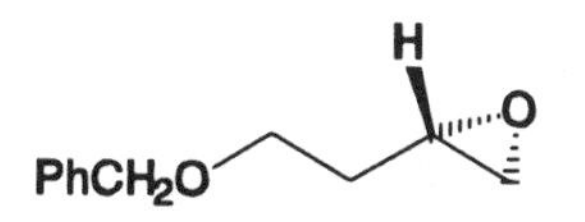

(The name 2-(phenylmethoxy)ethyl means that there is a $C_6H_5CH_2OCH_2CH_2$ substituent.)

o. Glattfeld, J. W. E.; Chittum, J. W. *J. Am. Chem. Soc.* **1933**, *55*, 3663

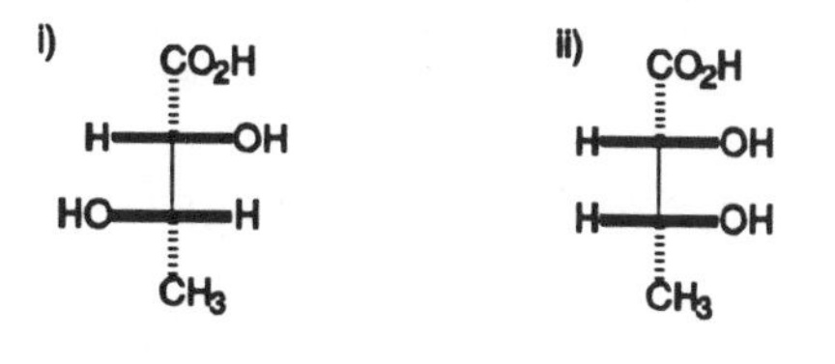

p. Chattopadhyay, S.; Mamdapur, V. R.; Chadha, M. S. *J. Chem. Res. (M)* **1990**, 1818.

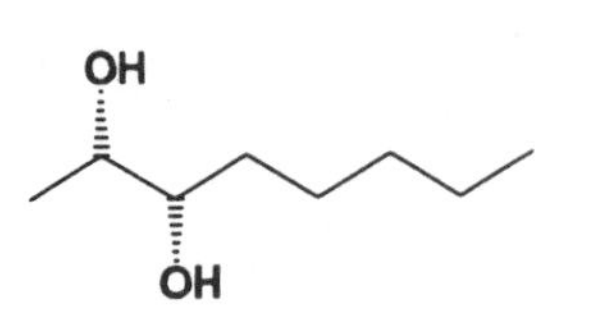

q. Hammarström, L.-G.; Berg, U.; Liljefors, T. *J. Chem. Res. (S)* **1990**, 152.

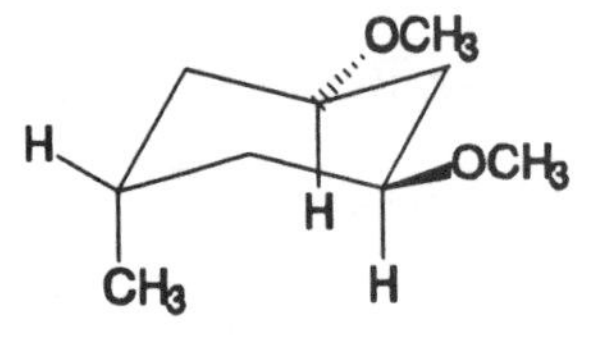

r. King, S. B.; Ganem, B. *J. Am. Chem. Soc.* **1994**, *116*, 562.

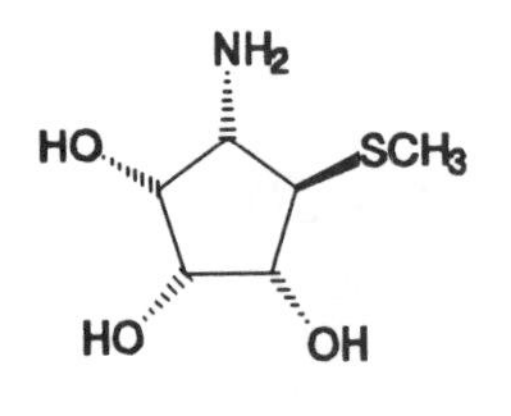

s. Moorthy, J. N.; Venkatesan, K. *J. Org. Chem.* **1991**, *56*, 6957.

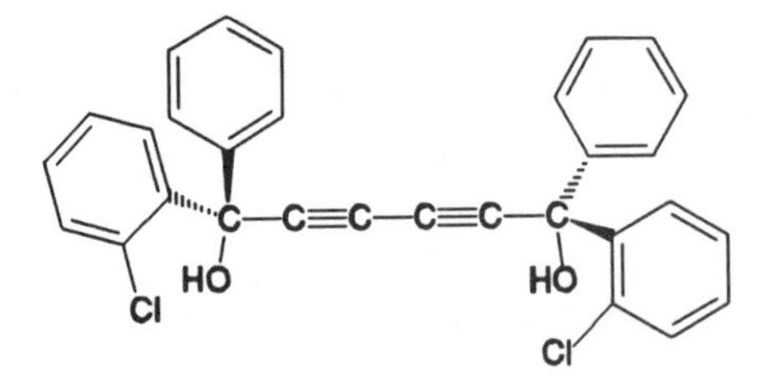

t. Andersen, K. K.; Colonna, S.; Stirling, C. J. M. *J. Chem. Soc., Chem. Commun.* **1973**, 645.

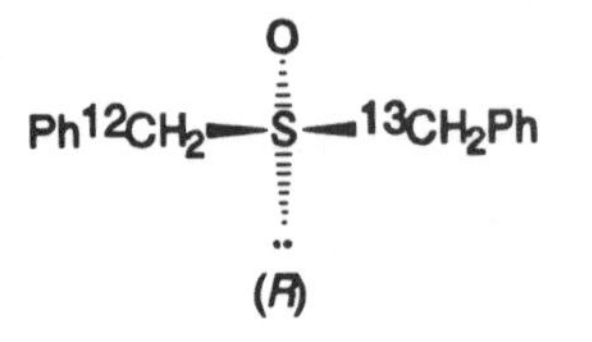

10. a. Coke, J. L.; Shue, R. S. *J. Org. Chem.* **1973**, *38*, 2210.

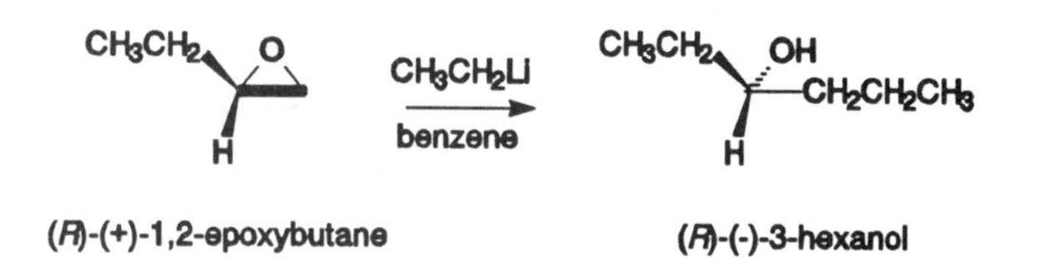

Retention.

b. Katsura, T.; Minamii, M. *Jpn. Kokai Tokkyo Hoho JP* 61,176,557; see *Chem. Abstr.* **1986**, *106*, 66799f; for a discussion, see Salaün, J. *Chem. Rev.* **1989**, *89*, 1247.

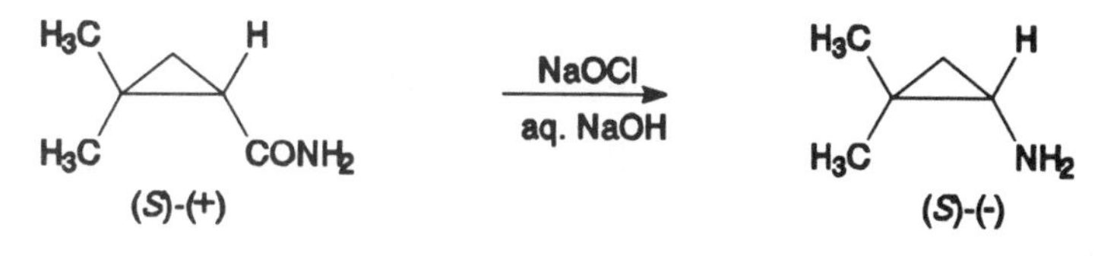

Retention.

c. Floss, H. G.; Lee, S. *Acc. Chem. Res.* **1993**, *26*, 116.

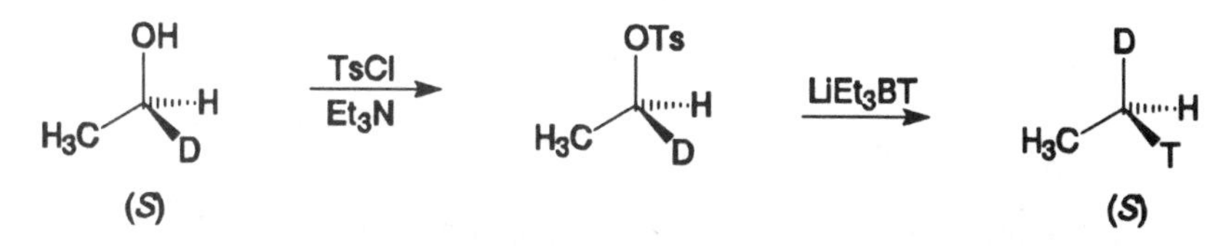

Inverson. See fig. 11, p. 121 in Floss, H. G.; Lee, S. *Acc. Chem. Res.* **1993**, *26*, 116.

d. Floss, H. G.; Lee, S. *Acc. Chem. Res.* **1993**, *26*, 116.

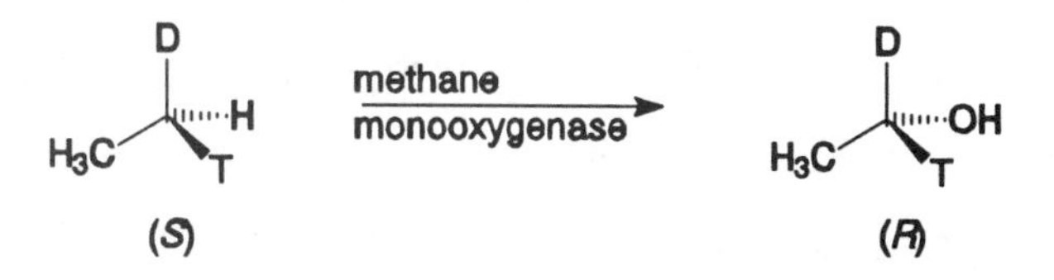

Retention. See fig. 11, p. 121 in Floss, H. G.; Lee, S. *Acc. Chem. Res.* **1993**, *26*, 116.

e. Walborsky, H. M.; Impastato, F. J.; Young, A. E. *J. Am. Chem. Soc.* **1964**, *86*, 3283.

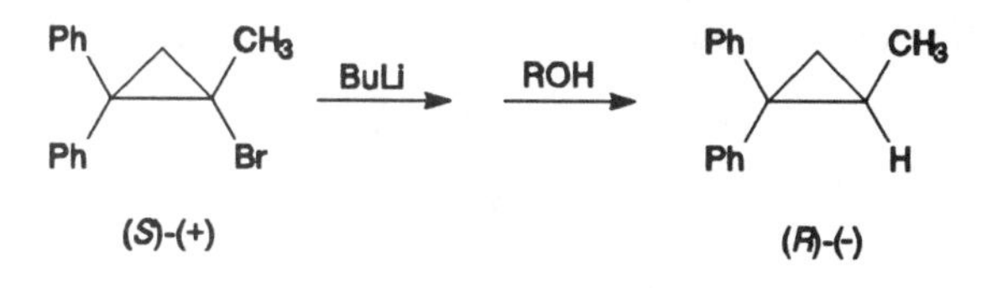

Retention.

f. Skell, P. S.; Pavlis, R. R.; Lewis, D. C.; Shea, K. J. *J. Am. Chem. Soc.* **1973**, *95*, 6735.

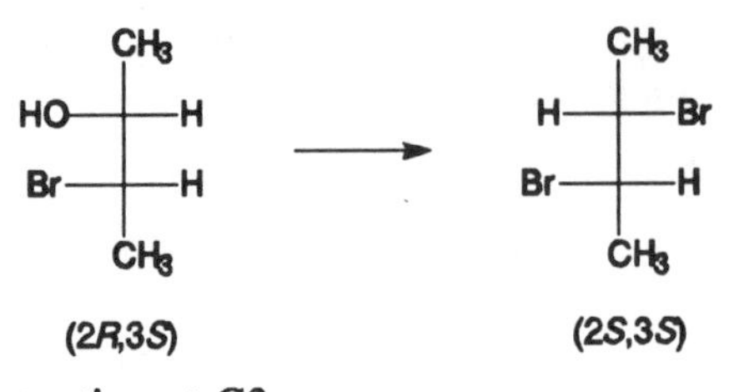

Inversion at C2 and retention at C3.

g. Wiberg, K. B. *J. Am. Chem. Soc.* **1952**, *74*, 3891.

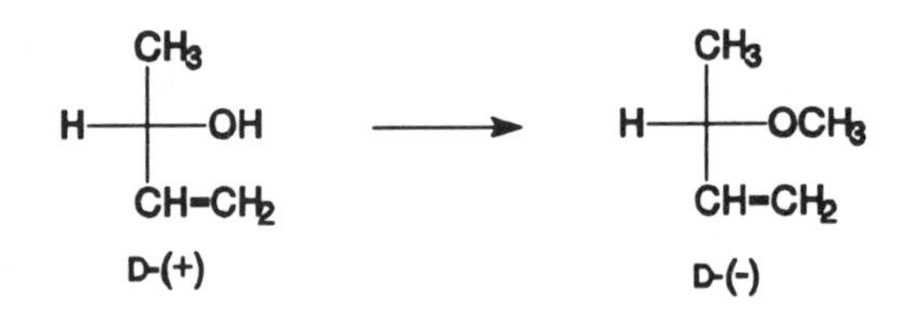

Retention

These problems require making a 3-D drawing of each structure and each product and analyzing for inversion or retention from the 3-D structures. Decisions cannot be made only on the basis of (*R*), (*S*), (+) or (-) designations.

11. Jennings, W. B. *Chem. Rev.* **1975**, *75*, 307.
Each of the four CH2 protons should have a different chemical shift. This problem is modified after one in the review by Jennings, W. B. *Chem. Rev.* **1975**, *75*, 307.

12. Carbon 3 in **109** (and **110**) is chirotopic because the molecule is chiral. Thus all atoms and spaces in the molecule are chirotopic. Carbon 3 is nonstereogenic because interchanging two groups attached to that carbon (easiest to see by interchanging the H and OH) does not generate a new stereoisomer; the interchange generates the same stereoisomer (with the Fischer projection turned upside down).

13. One approach is to make the individual substitutions; another is to determine the symmetry point group and apply the rules in the table.

a) barrelene belongs to point group D_{3h} so $p = 3$;

b) cubane belongs to point group D_{4h}, so $p = 3$;

c) bicyclo[1.1.0]butane belongs to point group C_{2v}, so $p = 2$. (Bicyclo[1.1.0]butane is not planar; if it were, it would belong to point group D_{2h} and would be (pro)^3-chiral.)

d) bicyclo[2.2.2]octane belongs to point group D_{3h}, so $p = 3$. It is interesting to note that replacing one of the bridgehead hydrogen atoms with another atom reduces the symmetry to C_{3v}. A structure belonging to point group D_{3h} is (pro)^3-chiral, and C_{3v} is (pro)^2-chiral. Replacing any other hydrogen atom in the molecule (with the exception of the remaining bridgehead hydrogen atom) would make the structure chiral in one step. However, replacing a carbon atom in a methylene group with ^{13}C (for example) reduces the symmetry to C_s, which is (pro)^1-chiral. Then another atom substitution can make the structure chiral (provided that the substitution further reduces the symmetry of the structure).

14. LeGoff, E.; Ulrich, S. E.; Denney, D. B. *J. Am. Chem. Soc.* **1958**, *80*, 622.

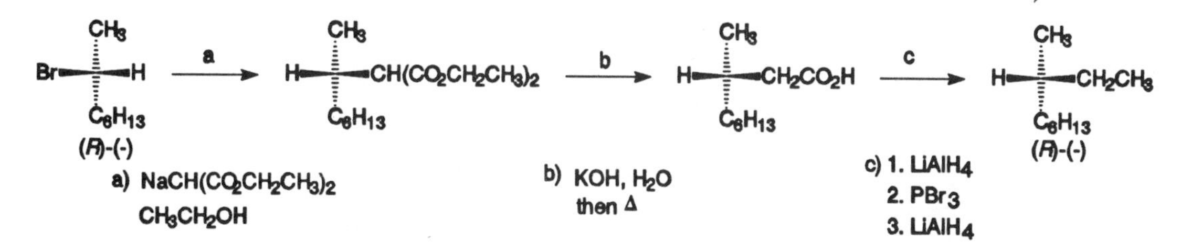

Reaction a is an S_N2 reaction, which proceeds with inversion of configuration. The fact that both the reactant and product of step a are (-) does not indicate retention. Step b is a hydrolysis and decarboxylation. Step c is a reduction of the carboxylic acid to an alcohol, conversion of the alcohol to a bromide, and reduction of the carbon-bromine bond to a carbon-hydrogen bond. Neither step b nor step c involve bond breaking at the chiral center, so all proceed with retention of configuration. Overall, therefore, the reaction sequence converts (*R*)-(-)-2-bromooctane to (*R*)-(-)-3-methylnonane with net inversion of configuration.

15. a) In the absence of isotopic labels, the parent reaction is only stereoselective, not stereospecific, because stereoisomeric reactants are not possible.

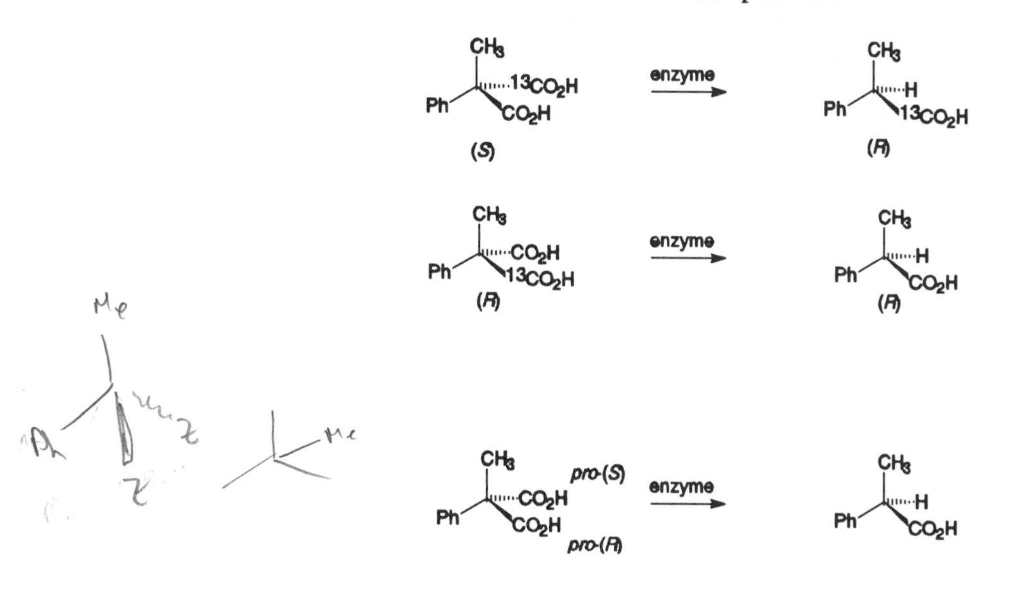

b) Miyamoto, K.; Tsuchiya, S.; Ohta, H. *J. Am. Chem. Soc.* **1992**, *114*, 6256.

The reaction takes place with inversion of configuration and with removal of the pro-(*R*) carboxyl group.

16. Cinquini, M.; Cozzi, F.; Sannicolò, F.; Sironi, A. *J. Am. Chem. Soc.* **1988**, *110*, 4363.

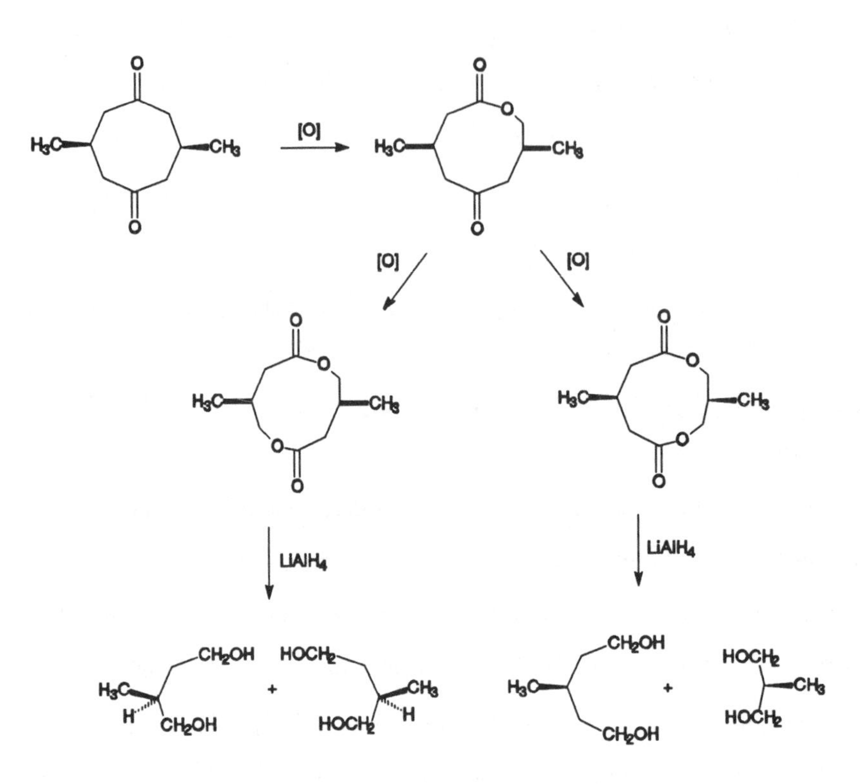

The second Baeyer-Villager oxidation converts the intermediate structure into either of two products. One of them is reduced with $LiAlH_4$ to two molecules of 2-methyl-1,4-butanediol. Reduction of the other product produces two achiral diols. See the reference cited for a more detailed discussion of the stereochemistry of these compounds.

17. Hoye, T. R.; Hanson, P. R.; Kovelesky, A. C.; Ocain, T. D.; Zhuang, Z. *J. Am. Chem. Soc.* **1991**, *113*, 9369

a.

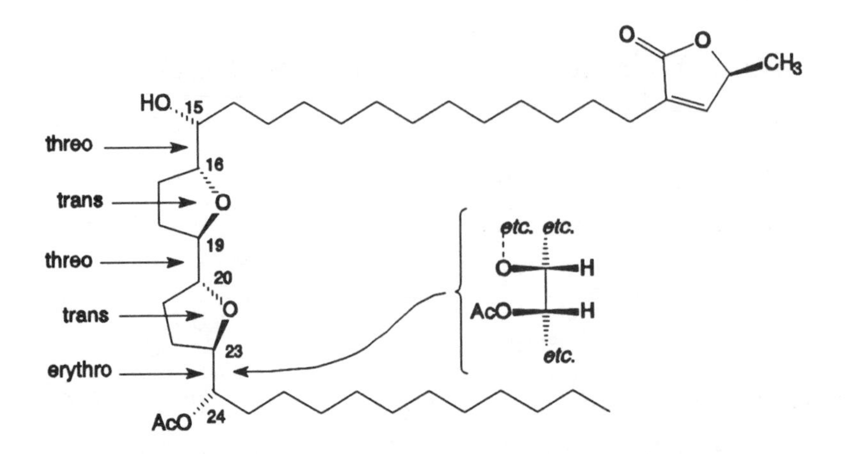

In order to determine the threo and erythro designations, it is helpful to redraw selected portions of the molecule as Fischer projections. One example is shown above.

b.

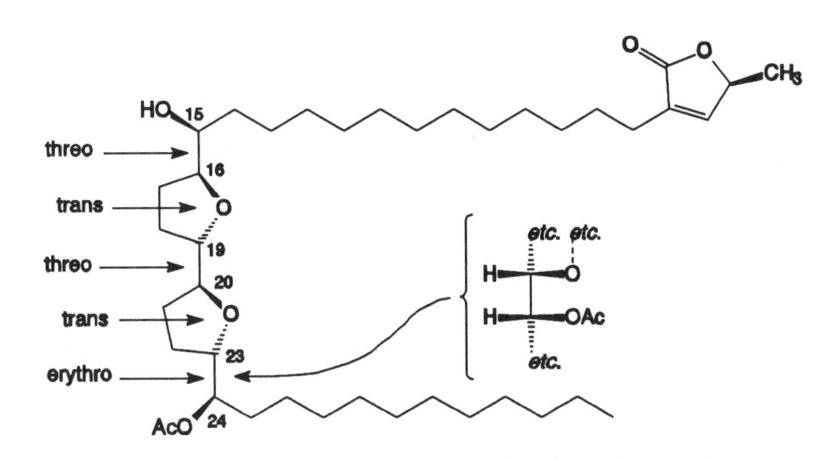

Note that the stereochemical designations do not change because the hexaepi structure is the mirror image of the original structure along the affected portions of the molecule. The hexaepi structure is not the enantiomer of the original structure, however, because the configuration of the chiral center in the lactone group is not changed.

18. There is no literature reference for this problem. The optical purity of the sample is -16.19/23.13 = .70 or 70%. That is also the enantiomeric excess. Since

$$(1\text{-}x) - x = 0.70$$

then $x = 0.15$. Therefore, the mole fraction of (-) enantiomer is 85%, while that of (+) enantiomer is 15%.

19. Polniaszek, R. P.; Dillard, L. W.; *Abstracts of the 203rd National Meeting of the American Chemical Society*, San Franciso, CA, April 5-10, 1992, Abstract ORGN 494.
Invert configuration at C2 (the carbon atom adjacent to the N) by moving the methyl group to the rear and the propyl group to the front.

20. Whitesides, G. M.; Kaplan, F.; Roberts, J. D. *J. Am. Chem. Soc.* **1963**, *85*, 2166.
Because of the adjacent chiral center, the two methylene protons are diastereotopic unless exchange at a rate that is fast on the NMR time scale averages their environments. A process such as reversible dissociation of the carbon-magnesium bond could lead to the observed results.

21. Streitwieser, Jr., A.; Granger, M. R. *J. Org. Chem.* **1967**, *32*, 1528.

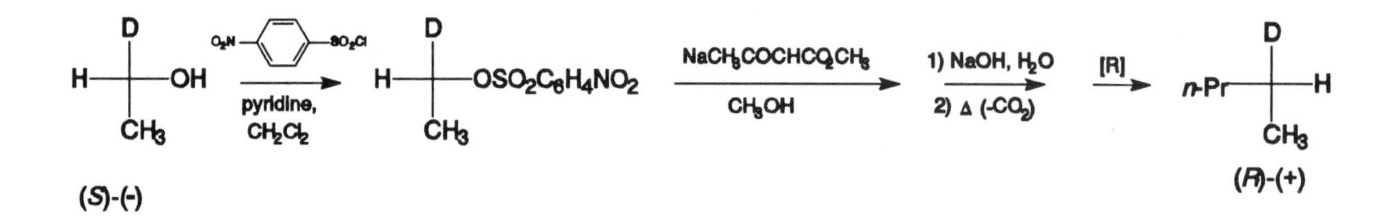

a) -0.123°/0.44 = -0.28°; b) (*S*). Since (-)-pentane-2-*d* has the (*S*) configuration, it must be the case that the product of this reaction sequence, (+)-pentane-2-*d* has the (*R*) configuration. The reaction of the *p*-nitrobenzenesulfonate with the sodium salt of methyl acetoacetate is an S_N2 reaction that occurs with inversion. All other reactions occur with retention of configuration at the chiral center.

22. Hilvert, D.; Nared, K. D. *J. Am. Chem. Soc.* **1988**, *110*, 5593.

a.

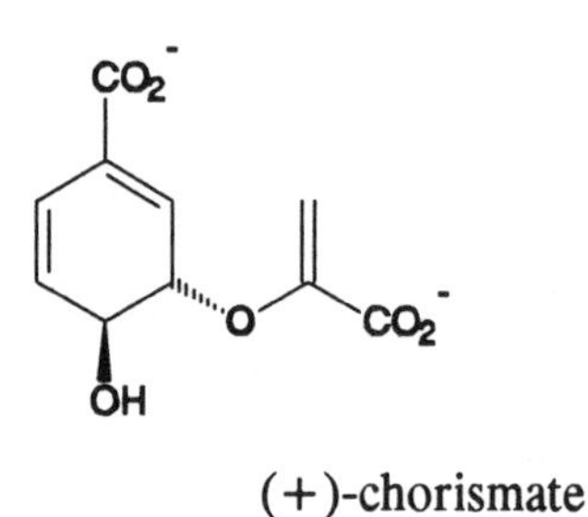

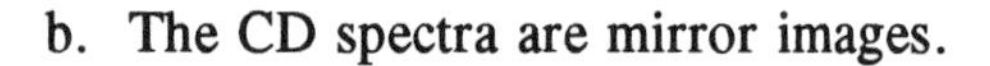
(+)-chorismate

b. The CD spectra are mirror images.

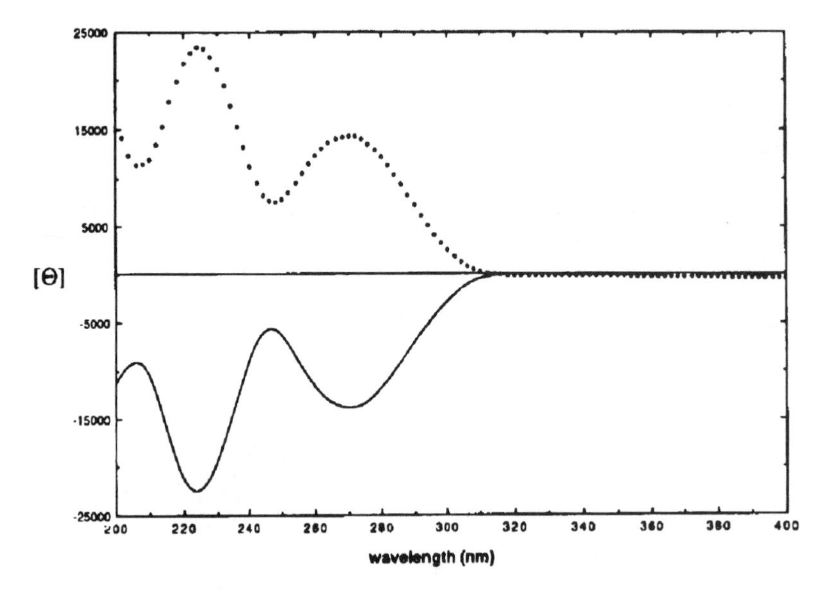

CD spectrum of (-)-chorismate (solid line) and (+)-chorismate (dotted line).

Chapter 3

1. Allinger, N. L.; Miller, M. A. *J. Am. Chem. Soc.* **1961**, *83*, 2145; Beckett, C. W.; Pitzer, K. S.; Spitzer, R. *J. Am. Chem. Soc.* **1947**, *69*, 2488.

   The cis isomer has two conformations, one in which both methyl groups are axial, and one in which both methyl groups are equatorial. The conformation with two axial methyl groups is 5.5 kcal/mol higher in energy than the e,e conformation.[12] Therefore, almost all molecules of *cis*-1,3-dimethylcyclohexane are in the e,e conformation, which has no butane *gauche* interactions. The trans isomer has one axial and one equatorial methyl group in either of two conformations. Therefore, the trans isomer should be higher in energy (and thus have a higher heat of combustion) than the cis isomer by 2 × 0.9 or 1.8 kcal/mol. The experimental value is 1.92 kcal/mol.[13]

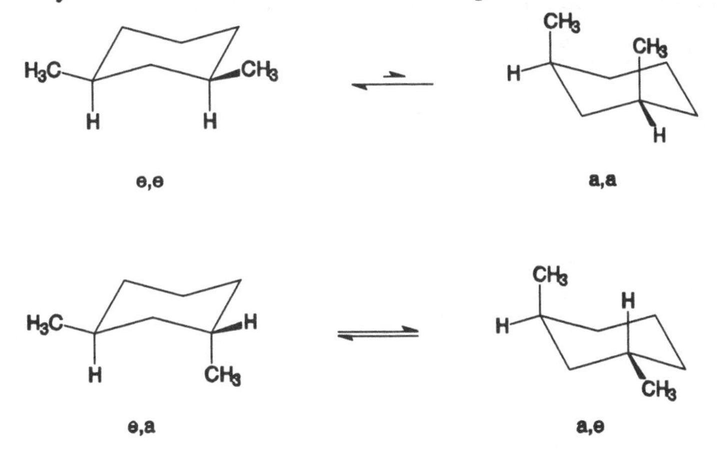

2. Eliel, E. L.; Allinger, N. L.; Angyal, S. J.; Morrison, G. A. *Conformational Analysis*, Wiley-Interscience: New York, 1965; p. 52 and references therein.

   In both cases the trans isomer is chiral and is capable of optical activity. For 1,2-dimethylcyclohexane, the trans isomer can exist predominantly in the diequatorial conformation, while the cis isomer must have one axial methyl group. Therefore the (chiral) trans isomer is the more stable isomer. In the case of 1,3-dimethylcyclohexane, it is achiral cis isomer that can have both methyl groups equatorial, so the chiral isomer is the less stable trans isomer.

3. Booth, H.; Everett, J. R. *J. Chem. Soc., Perkin Trans. 2* **1980**, 255.

   Only in *t*-butyl is there a forced enthalpy; in all other cases ethyl and isopropyl can adopt conformations in which the extra methyl group(s) point away from the cyclohexane ring; thus, their steric effect is not much greater than that of methyl. For t-butyl, however, no such conformation is possible, so the steric preference is much greater.

---

[12]Allinger, N. L.; Miller, M. A. *J. Am. Chem. Soc.* **1961**, *83*, 2145.

[13]Beckett, C. W.; Pitzer, K. S.; Spitzer, R. *J. Am. Chem. Soc.* **1947**, *69*, 2488.

4. Juaristi, E.; Labastida, V.; Antúnez, S. *J. Org. Chem.* **1991**, *56*, 4802.

   For the conversion of the axial benzyl-equatorial methyl conformer to the axial methyl-equatorial benzyl conformer, ΔS is +1.17 cal K^{-1} mol^{-1} and ΔH = +0.31 kcal/mol. Since $\Delta G = \Delta H\text{-}T\Delta S$, at low temperature, the equilibrium is dominated by the enthalpy term, while at higher temp the entropy term dominates.

5. a. Hydrogen bonding increases the effective size of the substituent.

   b. See Perrin, C. L.; Fabian, M. A. *Abstracts of the 209th National Meeting of the American Chemical Society*, Anaheim, CA, April 2-6, 1995, Abstract ORGN 6.

   The carboxylate ion leads to a structured solvent shell around the ionic carboxylate.

   c. Jensen, F. R.; Bushweller, C. H.; Beck, B. H. *J. Am. Chem. Soc.* **1969**, *91*, 344.

   Both the bond length and the polarizability of the halogen atom increases along the series F, Cl, Br, I. Both effects tend to reduce nonbonded interactions for axial halogen atoms.

6. Chupp, J. P.; Olin, J. F. *J. Org. Chem.* **1967**, *32*, 2297.

   Hydrogen bonding increases the effective size of the substituents on the bond about which rotation occurs, so rotation is slowed.

7. Huang, J.; Hedberg, K. *J. Am. Chem. Soc.* **1989**, *111*, 6909.

   a. The gauche conformer is stabilized both by the gauche effect and by hydrogen bonding.

   b. $\Delta G° = -\text{RT} \ln \text{K}$

   $K = .098/.902 = 0.109$

   $-RT \ln K = 2263 \text{ cal/mol} = 2.26 \text{ kcal/mol}$

   c. $\Delta H° = \Delta G° + T\Delta S° = 2263 + 513.15 \times .81 = 2678$ cal/mol = 2.7 kcal/mol. Technically, the calculation gives $\Delta U°$ (the internal energy), where $\Delta U° = \Delta H° + \Delta PV$, but the ΔPV term should be essentially 0 for interconversion of conformers.

8. For *n*-heptane: $2 \times (-10.08) + 5 \times (-4.95) = -44.91$. There are no butane gauche interactions. The literature value[14] is -44.88 kcal/mol.

   For 2-methylhexane: $3 \times (-10.08) + 3 \times (-4.95) + (1.90) = -46.99$ kcal/mol. There is one butane gauche interaction, so the corrected value is $-46.99 + 0.8 = -46.19$ kcal/mol. The experimental value is -46.59 kcal/mol.

   For 3-methylhexane, the count is as above but there are 2 butane gauche interactions, giving a net -45.39 kcal/mol. The literature value is -45.96 kcal/mol.

---

[14] Stull, D. R.; Westrum, Jr., E. F.; Sinke, G. C. *The Chemical Thermodynamics of Organic Compounds*; John Wiley & Sons: New York, 1969; pp. 249-252; also see Cox, J. D.; Pilcher, G. *Thermochemistry of Organic and Organometallic Compounds*; Academic Press: New York, 1970; p. 157 and references therein.

For 2,2-dimethylpentane, the count is 4 × (-10.05) + 2 × (-4.95) + 1 × (0.5) = -49.72 kcal/mol. There are 2 butane gauche interactions, so 1.6 kcal/mol is added, giving a total of -48.12 kcal/mol. The literature value is -49.27 kcal/mol.

For 2,3-dimethylpentane, the count is 4 × (-10.08) + 1 × (-4.95) + 2 × (-1.90) = -49.07. Now add 2.4 kcal/mol for 3 butane gauche interactions to get -46.67 kcal/mol. The literature value is -47.62.

For 3,3-dimethylpentane, the count is the same as for 2,2-dimethylpentane. Thus the strain free value is -49.72 kcal/mol. There are 4 butane gauche interactions, so we add +3.2 kcal/mol to get -46.52 kcal/mol. The experimental value is -48.17 kcal/mol.

Calculations for the other three isomers of C_7H_{16} can be carried out in similar fashion:

For 3-ethylpentane: 3 × (-10.08) + 3 × (-4.95) = 1 × (-1.90) = -46.99 kcal/mol. There are 3 butane gauche interactions, so the net is -44.59 kcal/mol. Literature is -45.33 kcal/mol.

For 2,4-dimethylpentane, 4 × (-10.08) + 1 × (-4.95) + 2 × (-1.90) = -49.07 kcal/mol. There are 2 butane gauche interactions, so the net is -47.47 kcal/mol. The literature value is -48.28 kcal/mol.

For 2,2,3-trimethylbutane, 5 × (-10.08) + 1 × (-1.9) + 1 × (0.5) = -51.8 kcal/mol. Four butane gauche interactions add 3.2, so the net is -48.6 kcal/mol. The literature value is -48.95 kcal/mol.

These data suggest that the heats of formations of a series of isomeric alkanes are influenced by the number of substituents (branches), the steric size of the substituents, and the proximity of the substituents to each other along the chain. Increasing the number of branches lowers the heat of formation as does increasing the size of the substituents on isomers with the same number of substituents. Substituents on adjacent carbon atoms interfere sterically more than do substituents on the same carbon atom, leading to a less negative heat of formation.

9. Twistane or tricyclo[4.4.0.$0^{3,8}$]decane: Whitlock, Jr., H. W. *J. Am. Chem. Soc.* **1962**, *84*, 3412.

Basketane or pentacyclo[4.4.0.$2^{2,5}$.$0^{3,8}$$0^{4,7}$]decane: Gassman, P. G.; Yamaguchi, R. *J. Org. Chem.* **1978**, *43*, 4654.

Cubane: Eaton, P. E.; Cole, Jr., T. W. *J. Am. Chem. Soc.* **1964**, *86*, 3157.

Fenestrane: Hoeve, W. T.; Wynberg, H. *J. Org. Chem.* **1980**, *45*, 2925.

Tetrahedrane: for a discussion, see Maier, G. *Angew. Chem., Int. Ed. Engl.* **1988**, *27*, 309 and references therein.

Tricyclo[2.1.0.$0^{1,3}$]pentane: Wiberg, K. B.; McMurdie, N.; McClusky, J. V.; Hadad, C. M. *J. Am. Chem. Soc.* **1993**, *115*, 10653.

10. Maier, G. *Angew. Chem., Int. Ed. Engl.* **1988**, *27*, 309.

The *t*-butyl groups block the attack of reagents (other than a proton) on the tetrahedrane skeleton. In addition, they provide a what has been called a "corset effect," which holds the molecule in a tetrahderal shape. Any reaction that involves distoration of the tetrahedrane framework would increase the steric barrier of the bulky substitents.

11. Turner, R. B.; Nettleton, Jr., D. E.; Perelman, M. *J. Am. Chem. Soc.* **1958**, *80*, 1430; Brown, H. C.; Berneis, H. L. *J. Am. Chem. Soc.* **1953**, *75*, 10; Saunders, Jr., W. H.; Cockerill, A. F. *Mechanisms of Elimination Reactions*; Wiley-Interscience: New York, 1973; p. 173.

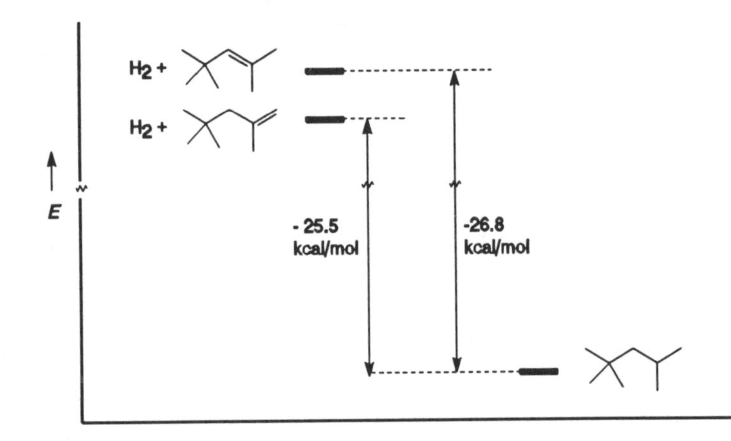

2,4,4-Trimethyl-1-pentene is less substituted. However, 2,4,4-trimethyl-2-pentene has a *t*-butyl group that is cis to a methyl group, which raises the energy of the compound due to van der Waals repulsion. The experimental data were reported by, and some discussion given by, Turner, R. B.; Nettleton, Jr., D. E.; Perelman, M. *J. Am. Chem. Soc.* **1958**, *80*, 1430. Brown, H. C.; Berneis, H. L. *J. Am. Chem. Soc.* **1953**, *75*, 10 had proposed the steric explanation for the greater stability of the less substituted isomer. For a discussion, see Saunders, Jr., W. H.; Cockerill, A. F. *Mechanisms of Elimination Reactions*; Wiley-Interscience: New York, 1973; p. 173.

12. Golan, O.; Goren, Z.; Biali, S. E. *J. Am. Chem. Soc.* **1990**, *112*, 9300; Juaristi, E.; Labastida, V.; Antúnez, S. *J. Org. Chem.* **1991**, *56*, 4802.

a) is *ap*; b) is +*sc*; c) is -*sc*; d) is -*sc*. In a) the two groups that determine the conformational designation are the CH_3 group on the upper carbon atom and the hydrogen of the lower carbon atom, because in each case these substituents are different from the other two substituents on the carbon atoms to which they are attached. In b), the two determining substituents are both hydrogens, for the same reasons. In (c and (d, the two groups are the hydrogen atom on the cyclohexane ring and the phenyl group on the $CH_2C_6H_5$ substituent.

13. Barton, D. H. R.; Cookson, R. C. *Quart. Rev. Chem. Soc.* **1956,** *10*, 44 (especially p. 58); Johnson, W. S. *J. Am. Chem. Soc.* **1953,** *75*, 1498.

    The trans isomer is more stable by 2.4 kcal/mol because of three butane gauche interactions.

14. Barton, D. H. R.; Cookson, R. C. *Quart. Rev. Chem. Soc.* **1956,** *10*, 44; Johnson, W. S. *J. Am. Chem. Soc.* **1953,** *75*, 1498.

    The *trans-anti-trans* isomer can have all chair conformations (with one gauche interaction), but the *trans-syn-trans* isomer has one cyclohexane ring in a boat conformation. Therefore the former isomer is more stable by at least 5.5 kcal/mol. Other steric interactions are also present in the *trans-syn-trans* isomer. A molecular mechanics calculation (PCMODEL) indicated that the *trans-syn-cis* isomer is more stable that the *trans-syn-trans* isomer by about 7.5 kcal/mol.

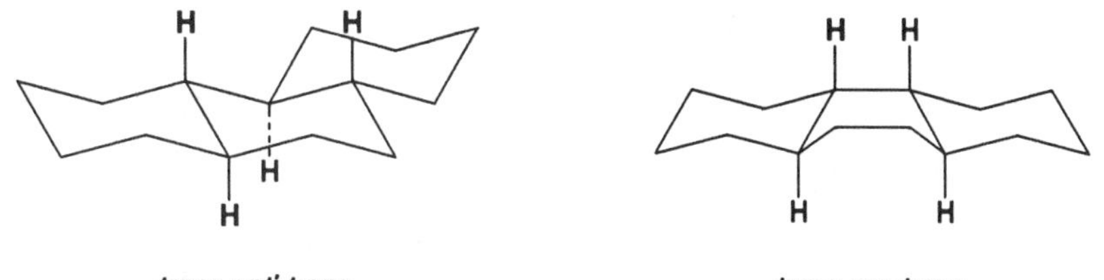

15. Jensen, F. R.; Bushweller, C. H.; Beck, B. H. *J. Am. Chem. Soc.* **1969**, *91*, 344.

    $\Delta G° = -RT \ln K$. In these calculations it is important to convert $\Delta G°$ values in kcal/mol to values in cal/mol, since the value of R often used (1.978 cal K^{-1} mol^{-1}) is in cal, not kcal. In addition, temperature must be converted to degrees K. The calculation yields $K = 6.95$. The distribution of equatorial and axial conformers is 87.4% and 12.6%, respectively.

16. Jensen, F. R.; Bushweller, C. H.; Beck, B. H. *J. Am. Chem. Soc.* **1969**, *91*, 344.

    $\Delta G° = 0.50$ kcal/mol.

17. Booth, H.; Everett, J. R. *J. Chem. Soc., Perkin Trans. 2* **1980**, 255.

    The entropy term dominates the calculation of $\Delta G°$ at 300° K; at 40° K the ΔH term dominates.

18. Eliel, E. L.; Manoharan, M. *J. Org. Chem.* **1981**, *46*, 1959.

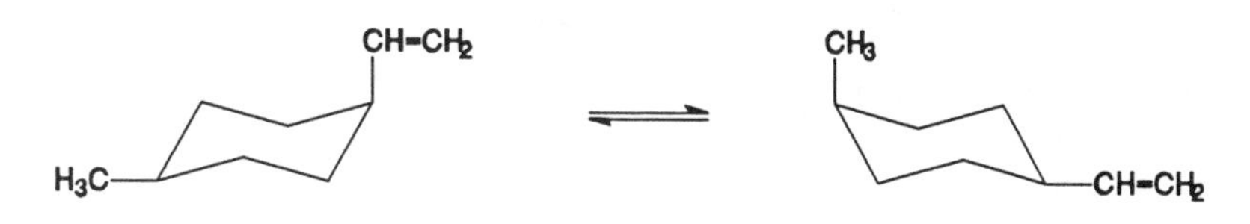

$$\Delta G° = A_{methyl} - A_{vinyl} = 0.06 \text{ kcal/mol}$$

Using 1.74 for the A value for the methyl group gives an A value for the vinyl group of 1.68.

19. There is no literature reference for this problem.

The chair conformation is the global minimum on the surface shown, while the half-chair conformation is the global maximum. (It would be reasonable to state that a global maximum is also a local maximum and that a global minimum is also a local minimum, since local maxima and minima are defined with regard to only a small portion of the potential energy surface without regard to the rest of the surface.) The twist boat conformations are local minima, while the boat conformation is a local maximum.

20. There is no literature reference for this problem.

Angle strain and torsional strain are easiest to identify in these structures, and there may also be some contributions from van der Waals strain and bond length deformation strain.

21. Kuhn, L. P. *J. Am. Chem. Soc.* **1958**, *80*, 5950.

The two OH groups must be gauche (and not anti) in order for intramolecular hydrogen bonding to be observed. In the meso diastereomer, the two R groups must also be gauche if the OH groups are gauche, so steric repulsion of the *t*-butyl groups makes the energy of this conformation prohibitive. In the racemic diastereomer the two R groups are anti when the two OH groups are gauche, so the size of the R groups does not affect intramolecular hydrogen bonding.

22. There is no literature reference for this problem.

a. Conformation B is puckered cyclobutane.

b. In comparing the planar conformation to the puckered conformation, the major differences are significantly greater angle strain but even more significantly less torsional strain in the puckered conformation. There is also slightly greater van der Waals strain in the puckered conformer. There is also somewhat greater angle strain in the puckered conformer, but this is offset somewhat by a smaller stretch-bend term. Overall the puckered conformation is more stable than the planar conformation by 0.92 kcal/mol.

23. There is no literature reference for this problem.

Conformer B has a boat conformation for the six-membered ring, while A has a chair conformation. The overall difference in steric energy is similar to the twist-boat - chair energy difference in cyclohexane.

Chapter 4

1. There is no literature reference for this problem.

   Substitute (α - ß) for (α + ß) in equations 4.18 and 4.19 (page 182) and solve.

2. There is no literature reference for this problem.

   The π bond orders are the same for all three species, because the only bonding between C1 and C2 and between C2 and C3 results from population of ψ_1. Therefore $P_{12} = P_{23} = 2 \times 0.500 \times 0.707 = 0.707$. Similarly, the free valence indices for the cation, radical and anion are all the same for each species. For C1 and C3, $\mathcal{F}_1 = \mathcal{F}_3 = 4.732 - (3 + 0.707) = 1.025$. $\mathcal{F}_2 = 4.732 - (3 + 0.707 + 0.707) = 0.318$.

3. There is no single literature reference for this problem. Hückel molecular orbital calculations for many compounds of interest are reported in Heilbronner und Straub, *Hückel Molecular Orbitals: HMO*; Springer-Verlag: New York, 1966. An HMO program is provided on a disk included with Rauk, A. *Orbital Interaction Theory of Organic Chemistry*; Wiley-Interscience: New York, 1994. The secular determinants are shown below in the solutions for problem 4.

4. There is no literature reference for this problem.

a) Hückel matrix for Cyclobutadiene

$$\begin{vmatrix} X & 1 & 0 & 1 \\ 1 & X & 1 & 0 \\ 0 & 1 & X & 1 \\ 1 & 0 & 1 & X \end{vmatrix} = 0$$

$\psi_4 = - 0.500\ \phi_1 + 0.500\ \phi_2 - 0.500\ \phi_3 + 0.500\ \phi_4$ $\quad E = \alpha - 2.000\ ß$

$\psi_3 = + 0.500\ \phi_1 - 0.500\ \phi_2 - 0.500\ \phi_3 + 0.500\ \phi_4$ $\quad E = \alpha$

$\psi_2 = - 0.500\ \phi_1 - 0.500\ \phi_2 + 0.500\ \phi_3 + 0.500\ \phi_4$ $\quad E = \alpha$

$\psi_1 = + 0.500\ \phi_1 + 0.500\ \phi_2 + 0.500\ \phi_3 + 0.500\ \phi_4$ $\quad E = \alpha + 2.000\ ß$

Note: reversing all signs in any wave function does not change the energy or bonding relationships. Thus, we may also write

$$\psi_2 = + 0.500\ \phi_1 + 0.500\ \phi_2 - 0.500\ \phi_3 - 0.500\ \phi_4$$

With 2, 1, 1, 0 electrons in the four MOs (populated from ψ_1 to ψ_4), the π energy is
$E_\pi = 4\ \alpha + 4.000\ ß$

Electron densities
$\rho_1 = 1.000$; $\rho_2 = 1.000$; $\rho_3 = 1.000$; $\rho_4 = 1.000$

Partial bond orders:
$P_{12} = P_{23} = P_{34} = P_{41} = 0.5000$

Free valence indices:
$\mathcal{F}_1 = .732$; $\mathcal{F}_2 = .732$; $\mathcal{F}_3 = .732$; $\mathcal{F}_4 = .732$

b) Hückel matrix for methylenecyclopropene

$$\begin{vmatrix} X & 1 & 0 & 0 \\ 1 & X & 1 & 1 \\ 0 & 1 & X & 1 \\ 0 & 1 & 1 & X \end{vmatrix} = 0$$

$\psi_4 = +\ 0.506\ \phi_1 - 0.749\ \phi_2 + 0.302\ \phi_3 + 0.302\ \phi_4$ $\quad E = \alpha - 1.481\ \beta$
$\psi_3 = +\ 0.000\ \phi_1 + 0.000\ \phi_2 - 0.707\ \phi_3 + 0.707\ \phi_4$ $\quad E = \alpha - 1.000\ \beta$
$\psi_2 = -\ 0.815\ \phi_1 - 0.254\ \phi_2 + 0.368\ \phi_3 + 0.368\ \phi_4$ $\quad E = \alpha + 0.311\ \beta$
$\psi_1 = +\ 0.282\ \phi_1 + 0.612\ \phi_2 + 0.523\ \phi_3 + 0.523\ \phi_4$ $\quad E = \alpha + 2.170\ \beta$

With 2, 2, 0, and 0 electrons in the four MOs, $E_\pi = 4\ \alpha + 4.962\ \beta$

Electron densities:
$\rho_1 = 1.4881$; $\rho_2 = 0.8768$; $\rho_3 = \rho_4 = 0.8176$

Partial bond orders:
$P_{12} = 0.7583$; $P_{23} = P_{24} = 0.4527$; $P_{34} = 0.8176$

Free valence indices:
$\mathcal{F}_1 = .974$; $\mathcal{F}_2 = .068$; $\mathcal{F}_3 = .462$; $\mathcal{F}_4 = .462$

c) Hückel matrix for fulvene

$$\begin{vmatrix} X & 1 & 0 & 0 & 1 & 1 \\ 1 & X & 1 & 0 & 0 & 0 \\ 0 & 1 & X & 1 & 0 & 0 \\ 0 & 0 & 1 & X & 1 & 0 \\ 1 & 0 & 0 & 1 & X & 0 \\ 1 & 0 & 0 & 0 & 0 & X \end{vmatrix} = 0$$

The resulting MOs and energy levels are shown in compact form in the following table. For example,

$$\psi_1 = 0.523\ \phi_1 + 0.429\ \phi_2 + 0.385\ \phi_3 + 0.385\ \phi_4 + 0.429\ \phi_5 + 0.\ 247\ \phi_6$$

and the energy of ψ_1 is $\alpha + 2.115\ \beta$

	MO 1	MO 2	MO 3	MO 4	MO 5	MO 6
c\E	2.115	1.000	.618	-.254	-1.618	-1.861
1	.523	-.500	.000	.190	.000	-.664
2	.429	.000	-.602	.351	-.372	.439
3	.385	.500	-.372	-.280	.602	-.153
4	.385	.500	.372	-.280	-.602	-.153
5	.429	.000	.602	.351	.372	.439
6	.247	-.500	.000	-.749	.000	.357

With 2, 2, 2, 0, 0, and 0 electrons in the 6 MOs, $E_\pi = 6\ \alpha + 7.466\ \beta$

Values of ρ_i (in bold) and P_{ij} are summarized in the following table:

	1	2	3	4	5	6
1	**1.0470**	.4491			.4491	.7586
2	.4491	**1.0923**	.7779			
3		.7779	**1.0730**	.5202		
4			.5202	**1.0730**	.7779	
5	.4491			.7779	**1.0923**	
6	.7586					**.6223**

For example, ρ_1 = 1.0470 and P_{15} = 0.4491.

Free valence indices:
$\mathcal{F}_1$ = .075; $\mathcal{F}_2$ = .505; $\mathcal{F}_3$ = .434; $\mathcal{F}_4$ = .434; $\mathcal{F}_5$ = .505; $\mathcal{F}_6$ = .973

d) Hückel matrix for styrene

$$\begin{vmatrix} X & 1 & 0 & 0 & 0 & 1 & 1 & 0 \\ 1 & X & 1 & 0 & 0 & 0 & 0 & 0 \\ 0 & 1 & X & 1 & 0 & 0 & 0 & 0 \\ 0 & 0 & 1 & X & 1 & 0 & 0 & 0 \\ 0 & 0 & 0 & 1 & X & 1 & 0 & 0 \\ 1 & 0 & 0 & 0 & 1 & X & 0 & 0 \\ 1 & 0 & 0 & 0 & 0 & 0 & X & 1 \\ 0 & 0 & 0 & 0 & 0 & 0 & 1 & X \end{vmatrix} = 0$$

Table of molecular orbitals and energy levels:

	MO 1	MO 2	MO 3	MO 4	MO 5	MO 6	MO 7	MO 8
c\E	2.136	1.414	1.000	.662	-.662	-1.000	-1.414	-2.136
1	.513	-.354	.000	-.334	-.334	.000	-.354	.513
2	.394	.000	-.500	-.308	.308	-.500	.000	-.394
3	.329	.354	-.500	.130	.130	.500	.354	.329
4	.308	.500	.000	.394	-.394	.000	-.500	-.308
5	.329	.354	.500	.130	.130	-.500	.354	.329
6	.394	.000	.500	-.308	.308	.500	.000	-.394
7	.308	-.500	.000	.394	-.394	.000	.500	-.308
8	.144	-.354	.000	.595	.595	.000	-.354	.144

With 2, 2, 2, 2, 0, 0, 0, and 0 in the 8 MOs, $E_\pi = 8\,\alpha + 10.424\,\beta$

Values of ρ_i (in bold) and P_{ij} are summarized in the following table:

	1	2	3	4	5	6	7	8
1	**1.0000**	.6101				.6101	.4059	
2	.6101	**1.0000**	.6787					
3		.6787	**1.0000**	.6586				
4			.6586	**1.0000**	.6586			
5				.6586	**1.0000**	.6787		
6	.6101				.6787	**1.0000**		
7	.4059						**1.0000**	.9113
8							.9113	**1.0000**

Free valence indices:

$\mathcal{F}_1$ = .106; $\mathcal{F}_2$ = .443; $\mathcal{F}_3$ = .395; $\mathcal{F}_4$ = .415; $\mathcal{F}_5$ = .395; $\mathcal{F}_6$ = .443;
$\mathcal{F}_7$= .415; $\mathcal{F}_8$ = .821

e) Hückel matrix for naphthalene

$$\begin{vmatrix} X & 1 & 0 & 0 & 0 & 0 & 0 & 0 & 1 & 0 \\ 1 & X & 1 & 0 & 0 & 0 & 0 & 0 & 0 & 0 \\ 0 & 1 & X & 1 & 0 & 0 & 0 & 0 & 0 & 0 \\ 0 & 0 & 1 & X & 0 & 0 & 0 & 0 & 0 & 1 \\ 0 & 0 & 0 & 0 & X & 1 & 0 & 0 & 0 & 1 \\ 0 & 0 & 0 & 0 & 1 & X & 1 & 0 & 0 & 0 \\ 0 & 0 & 0 & 0 & 0 & 1 & X & 1 & 0 & 0 \\ 0 & 0 & 0 & 0 & 0 & 0 & 1 & X & 1 & 0 \\ 1 & 0 & 0 & 0 & 0 & 0 & 0 & 1 & X & 1 \\ 0 & 0 & 0 & 1 & 1 & 0 & 0 & 0 & 1 & X \end{vmatrix} = 0$$

Table of molecular orbitals and energy levels:

	MO 1	MO 2	MO 3	MO 4	MO 5	MO 6	MO 7	MO 8	MO 9	MO10
c\E	2.303	1.618	1.303	1.000	.618	-.618	-1.000	-1.303	-1.618	-2.303
1	.301	-.263	.400	.000	-.425	-.425	.000	.400	.263	.301
2	.231	-.425	.174	.408	-.263	.263	.408	-.174	-.425	-.231
3	.231	-.425	-.174	.408	.263	.263	-.408	-.174	.425	.231
4	.301	-.263	-.400	.000	.425	-.425	.000	.400	-.263	-.301
5	.301	.263	-.400	.000	-.425	.425	.000	.400	.263	-.301
6	.231	.425	-.174	.408	-.263	-.263	-.408	-.174	-.425	.231
7	.231	.425	.174	.408	.263	-.263	.408	-.174	.425	-.231
8	.301	.263	.400	.000	.425	.425	.000	.400	-.263	.301
9	.461	.000	.347	-.408	.000	.000	-.408	-.347	.000	-.461
10	.461	.000	-.347	-.408	.000	.000	.408	-.347	.000	.461

With 2, 2, 2, 2, 2, 0, 0, 0, 0, and 0 in the ten MOs, $E_\pi = 10\,\alpha + 13.683\,\beta$

Values of ρ_i (in bold) and P_{ij} are summarized in the following table:

	1	2	3	4	5	6	7	8	9	10
1	**1.0000**	.7246							.5547	
2	.7246	**1.0000**	.6032							
3		.6032	**1.0000**	.7246						
4			.7246	**1.0000**						.5547
5					**1.0000**	.7246				.5547
6					.7246	**1.0000**	.6032			
7						.6032	**1.0000**	.7246		
8							.7246	**1.0000**	.5547	
9	.5547							.5547	**1.0000**	.5182
10				.5547	.5547				.5182	**1.0000**

Free valence indices:

$\mathcal{F}_1$ = .453; $\mathcal{F}_2$ = .404; $\mathcal{F}_3$ = .404; $\mathcal{F}_4$ = .453; $\mathcal{F}_5$ = .453; $\mathcal{F}_6$ = .404;
$\mathcal{F}_7$ = .404; $\mathcal{F}_8$ = .453; $\mathcal{F}_9$ = .104; $\mathcal{F}_{10}$ = .104

f) Hückel matrix for biphenylene

$$\begin{vmatrix}
X & 1 & 0 & 0 & 0 & 1 & 0 & 0 & 0 & 0 & 0 & 1 \\
1 & X & 1 & 0 & 0 & 0 & 0 & 0 & 0 & 0 & 0 & 0 \\
0 & 1 & X & 1 & 0 & 0 & 0 & 0 & 0 & 0 & 0 & 0 \\
0 & 0 & 1 & X & 1 & 0 & 0 & 0 & 0 & 0 & 0 & 0 \\
0 & 0 & 0 & 1 & X & 1 & 0 & 0 & 0 & 0 & 0 & 0 \\
1 & 0 & 0 & 0 & 1 & X & 1 & 0 & 0 & 0 & 0 & 0 \\
0 & 0 & 0 & 0 & 0 & 1 & X & 1 & 0 & 0 & 0 & 1 \\
0 & 0 & 0 & 0 & 0 & 0 & 1 & X & 1 & 0 & 0 & 0 \\
0 & 0 & 0 & 0 & 0 & 0 & 0 & 1 & X & 1 & 0 & 0 \\
0 & 0 & 0 & 0 & 0 & 0 & 0 & 0 & 1 & X & 1 & 0 \\
0 & 0 & 0 & 0 & 0 & 0 & 0 & 0 & 0 & 1 & X & 1 \\
1 & 0 & 0 & 0 & 0 & 0 & 1 & 0 & 0 & 0 & 1 & X
\end{vmatrix} = 0$$

Table of molecular orbitals and energy levels:

	MO 1	MO 2	MO 3	MO 4	MO 5	MO 6	MO 7	MO 8	MO 9	MO10
c\E	2.532	1.802	1.347	1.247	.879	.445	-.445	-.879	-1.247	-1.347
1	.422	.164	-.225	.296	.147	-.368	.368	.147	-.296	.225
2	.225	.296	.147	.368	.422	-.164	-.164	-.422	.368	.147
3	.147	.368	.422	.164	.225	.296	-.296	.225	-.164	-.422
4	.147	.368	.422	-.164	-.225	.296	.296	.225	-.164	.422
5	.225	.296	.147	-.368	-.422	-.164	.164	-.422	.368	-.147
6	.422	.164	-.225	-.296	-.147	-.368	-.368	.147	-.296	-.225
7	.422	-.164	-.225	-.296	.147	.368	-.368	.147	.296	.225
8	.225	-.296	.147	-.368	.422	.164	.164	-.422	-.368	.147
9	.147	-.368	.422	-.164	.225	-.296	.296	.225	.164	-.422
10	.147	-.368	.422	.164	-.225	-.296	-.296	.225	.164	.422
11	.225	-.296	.147	.368	-.422	.164	-.164	-.422	-.368	-.147
12	.422	-.164	-.225	.296	-.147	.368	.368	.147	.296	-.225

	MO11	MO12
	-1.802	-2.532
1	.164	-.422
2	-.296	.225
3	.368	-.147
4	-.368	.147
5	.296	-.225
6	-.164	.422
7	-.164	-.422
8	.296	.225
9	-.368	-.147
10	.368	.147
11	-.296	-.225
12	.164	.422

With 2, 2, 2, 2, 2, 2, 0, 0, 0, 0, 0, and 0 electrons in the 12 MOs, $E_\pi = 12\,\alpha + 16.505\,\beta$

Values of ρ_i (in bold) and P_{ij} are summarized in the following table:

	1	2	3	4	5	6	7	8	9	10	11	12
1	**1.0000**	.6830				.5648						.2634
2	.6830	**1.0000**	.6208									
3		.6208	**1.0000**	.6907								
4			.6907	**1.0000**	.6208							
5				.6208	**1.0000**	.6830						
6	.5648				.6830	**1.0000**	.2634					
7						.2634	**1.0000**	.6830				.5648
8							.6830	**1.0000**	.6208			
9								.6208	**1.0000**	.6907		
10									.6907	**1.0000**	.6208	
11										.6208	**1.0000**	.6830
12	.2634						.5648				.6830	**1.0000**

Free valence indices:
$\mathcal{F}_1 = .221$; $\mathcal{F}_2 = .428$; $\mathcal{F}_3 = .420$; $\mathcal{F}_4 = .420$; $\mathcal{F}_5 = .428$; $\mathcal{F}_6 = .221$;
$\mathcal{F}_7 = .221$; $\mathcal{F}_8 = .428$; $\mathcal{F}_9 = .420$, $\mathcal{F}_{10} = .420$; $\mathcal{F}_{11} = .428$; $\mathcal{F}_{12} = .221$

5. There is no literature reference for this problem. Answers are given in the answers to question 4.

6. The orbital symmetries are indicated schematically. A plus sign means that the coefficient is positive for a given atomic orbital; a minus sign means that the coefficient is negative; a zero indicates a node at a particular position.

heptatrienyl

ψ_7	+	−	+	−	+	−	+	
ψ_6	+	−	+	0	−	+	−	
ψ_5	+	−	−	+	−	−	+	
ψ_4	+	0	−	0	+	0	−	(NBMO)
ψ_3	+	+	−	−	−	+	+	
ψ_2	+	+	+	0	−	−	−	
ψ_1	+	+	+	+	+	+	+	

octatetraene

ψ_8	+	−	+	−	+	−	+	−
ψ_7	+	−	+	−	−	+	−	+
ψ_6	+	−	0	+	−	0	+	−
ψ_5	−	+	+	−	−	+	+	−
ψ_4	+	+	−	−	+	+	−	−
ψ_3	+	+	0	−	−	0	+	+
ψ_2	+	+	+	+	−	−	−	−
ψ_1	+	+	+	+	+	+	+	+

nonatetraenyl

ψ_9	+	−	+	−	+	−	+	−	+	
ψ_8	+	−	+	−	0	+	−	+	−	
ψ_7	+	−	+	+	−	+	+	−	+	
ψ_6	+	−	−	+	0	−	+	+	−	
ψ_5	+	0	−	0	+	0	−	0	−	(NBMO)
ψ_4	+	+	−	−	0	+	+	−	−	
ψ_3	+	+	+	−	−	−	+	+	+	
ψ_2	+	+	+	+	0	−	−	−	−	
ψ_1	+	+	+	+	+	+	+	+	+	

decapentaene

ψ_{10}	+	−	+	−	+	−	+	−	+	−
ψ_9	+	−	+	−	+	+	−	+	−	+
ψ_8	+	−	+	+	−	+	−	−	+	−
ψ_7	−	+	+	−	+	+	−	+	+	−
ψ_6	+	−	−	+	+	−	−	+	+	−
ψ_5	+	+	−	−	+	+	−	−	+	+
ψ_4	+	+	−	−	−	+	+	+	−	−
ψ_3	+	+	+	−	−	−	−	+	+	+
ψ_2	+	+	+	+	+	−	−	−	−	−
ψ_1	+	+	+	+	+	+	+	+	+	+

Note: the exact placement of the nodes may vary, but the important factors are i) the proper number of nodes, ii) the symmetry of the placement of the nodes, and ii) approximately increasing energy as one goes from ψ_n to ψ_{n+1}.

7. a) is odd alternant (OA). The others are EA, NA, and OA.

8. To solve the problem as stated, set up the equations:

$$c_1 + c_3 + c_5 = 0$$

$$c_3 + c_5 = 0$$

By subtracting the second equation from the first, it is immediately apparent that $c_1 = 0$. This result is initially surprising, because the radical character on carbon 1 is 0, even though we are calculating properties of a structure with a full radical shown at that position. This result is somewhat easier to comprehend if we note that resonance (VB) theory would predict an exocyclic double bond to a cyclobutenyl radical as the more stable structure, so there really should not be a lot of radical character on C1.

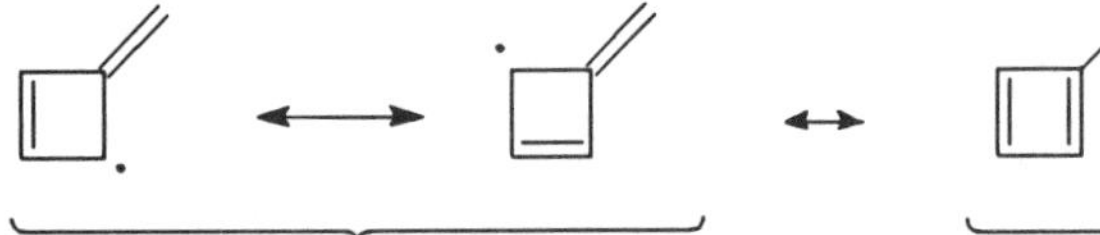

Major contributors to resonance hybrid

Negligible contributors to resonance hybrid due to antiaromaticity of cyclobutadiene

To verify this result with an HMO calculation:

Input Hückel matrix

$$\begin{vmatrix} X & 1 & 0 & 0 & 0 \\ 1 & X & 1 & 0 & 1 \\ 0 & 1 & X & 1 & 0 \\ 0 & 0 & 1 & X & 1 \\ 0 & 1 & 0 & 1 & X \end{vmatrix} = 0$$

Table of molecular orbitals and energy levels:

	MO 1	MO 2	MO 3	MO 4	MO 5
c\E	2.136	.662	.000	-.662	-2.136
1	.261	-.657	.000	-.657	.261
2	.557	-.435	.000	.435	-.557
3	.465	.185	-.707	.185	.465
4	.435	.557	.000	-.557	-.435
5	.465	.185	.707	.185	.465

With 2, 2, 1, 0, and 0 electrons in the 5 MOs, $E_\pi = 5\,\alpha + 5.596\,\beta$

Values of ρ_i (in bold) and P_{ij} are summarized in the following table:

	1	2	3	4	5
1	**1.0000**	.8629			
2	.8629	**1.0000**	.3574		.3574
3		.3574	**1.0000**	.6101	
4			.6101	**1.0000**	.6101
5		.3574		.6101	**1.0000**

Free valence indices:

$\mathcal{F}_1 = .869$; $\mathcal{F}_2 = .154$; $\mathcal{F}_3 = .764$; $\mathcal{F}_4 = .512$; $\mathcal{F}_5 = .764$

9. Methylenecyclopropene was synthesized by Billups, W. E.; Lin, L.-J.; Casserly, E. W. *J. Am. Chem. Soc.* **1984,** *106*, 3698 and by Staley, S. W.; Norden, T. D. *J. Am. Chem. Soc.* **1984,** *106*, 3699. For a discussion of the structure of methylenecyclopropene (dipole moment of 1.90 D) and a theoretical description that is more complex than the simple resonance description, see by Norden, T. D.; Staley, S. W.; Taylor, W. H.; Harmony, M. D. *J. Am. Chem. Soc.* **1986,** *108*, 7912. Also see Bachrach, S. M. *J. Org. Chem.* **1990,** *55*, 4961.

   Methylenecyclopropene was found to have charges; naphthalene was not. Yes. Resonance theory would have predicted a resonance hybrid with contribution from resonance structures with an aromatic cyclopropenyl cation bonded to a methylene anion substituent, so the hybrid would be quite polar.

10. For experimental data, see Staley, S. W.; Norden, T. D.; Taylor, W. H.; Harmony, M. D. *J. Am. Chem. Soc.* **1987**, *109*, 7641; also see Bachrach, S. M. *J. Org. Chem.* **1990,** *55*, 4961. The argument is just the same as that for methylenecyclopropene: the stability of the aromatic cyclopropenyl cation makes the polar form more stable. In addition, the greater electronegativity of oxygen than carbon also enhances the polarity of the molecule.

11. a) The three resonance structures for naphthalene are the following:

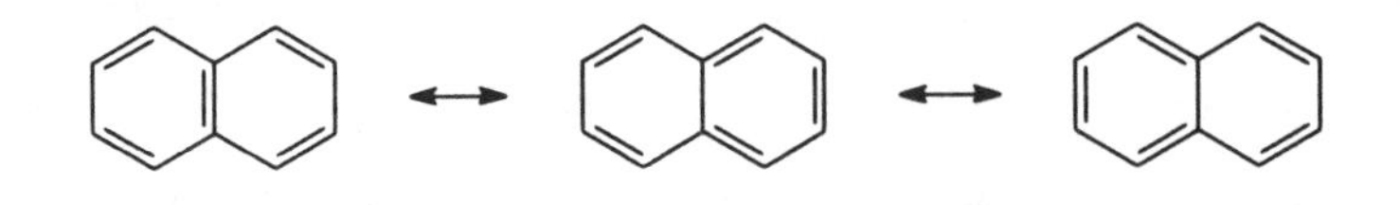

The five resonance structures of phenanthrene are the following:

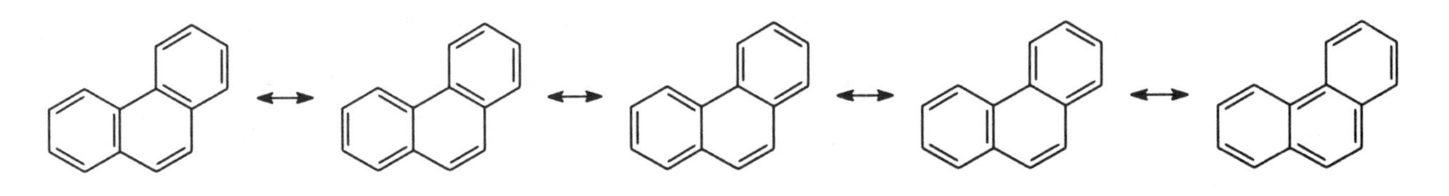

These results can be confirmed by application of the vertex deletion method to each compound. For naphthalene:

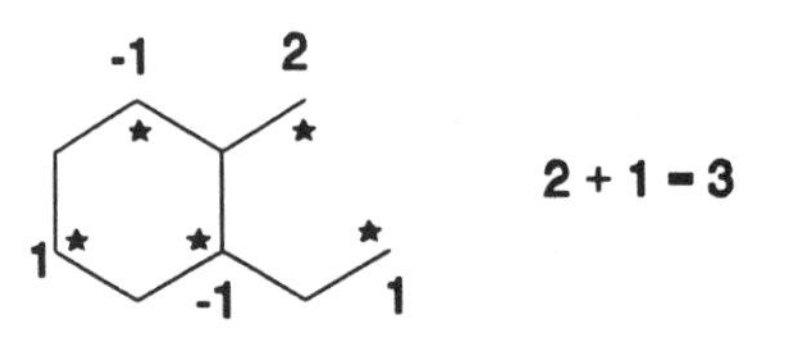

For phenanthrene:

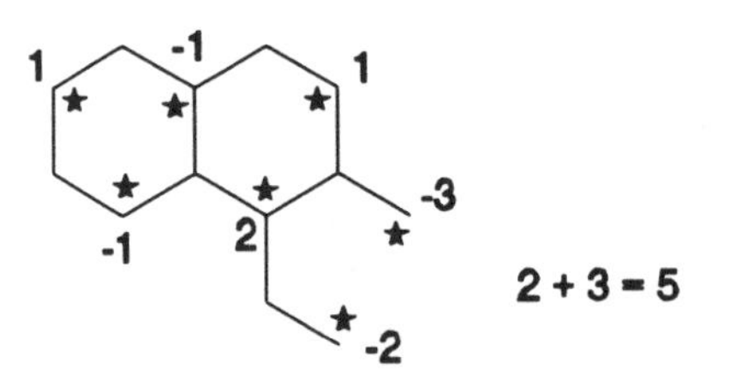

b) In phenanthrene, 4 of 5 resonance structures have double bonds between carbon atoms 9 and 10, so this bond is rather similar to an olefinic double bond. Furthermore, electrophilic addition across the 9,10 bond would give a product that retains four of the five resonance structures of the parent compound. In naphthalene there is a double bond between carbon atoms 1 and 2 in two of the three resonance structures, so the molecule is relatively less olefinic between these two carbon atoms.

12. Theoretical calculations of the structure of the allyl cation, radical and anion were given by Schleyer, P. v. R. *J. Am. Chem. Soc.* **1985**, *107*, 4793. The lengths of the carbon-carbon bonds were found to be similar in all three compounds, consistent with the prediction of HMO theory. The C-C-C bond angle was found to increase along the series allyl cation, radical, anion.

Vajda, E.; Tremmel, J.; Rozsondai, B.; Hargittai, I.; Maltsev, A. K.; Kagramanov, N. D.; Nefedov, O. M. *J. Am. Chem. Soc.* **1986**, *108*, 4352 determined the geometry of the allyl radical by using high-temperature electron diffraction. The results indicated a planar geometry, C-C bond length of 1.428 Å, and a C-C-C bond angle of 124.6 $\pm$ 3.4°.

Dorigo, A. E.; Li, Y.; Houk, K. N. *J. Am. Chem. Soc.* **1989**, *111*, 6942, reported that the calculated rotational barriers for the benzyl cation, radical and anion are 45.4, 20.0 and 28.1 kcal/mol, respectively, while the rotational barriers for the allyl cation, radical and anion are 34.9, 14.1 and 19.0 kcal/mol, respectively. This paper also cited experimental data for the allyl species. Korth, H.-G.; Trill, H.; Sustmann, R. *J. Am. Chem. Soc.* **1981**, *103*, 4483, determined a rotational barrier for the allyl radical of 15.7 $\pm$ 1.0 kcal/mol; therefore value of delocalization energy for allyl was found to be 14.0 - 14.5 kcal/mol. For the allyl anion, an experimental value for rotation about the C-1 - C-2 bond of allylcesium was found by Thompson, T. B.; Ford, W. T. *J. Am. Chem. Soc.* **1979**, *101*, 5459, to be 18.0 kcal/mol. This represents a lower limit for the rotational barrier about an allyl anion. A semi-empirical value (based on a correlation of calculated and experimental values for other systems) for the rotation about the C-1 - C-2 bond of the allyl cation in solution is 23.7 kcal/mol: Mayr, H.; Förner, W.; Schleyer, P. v. R. *J. Am. Chem. Soc.* **1979**, *101*, 6032. Gobbi, A.; Frenking, G. *J. Am. Chem. Soc.* **1994**, *116*, 9275 reported calculated rotational barriers for the allyl cation, radical and anion.

Krusic, P. J.; Meakin, P.; Smart, B. E. *J. Am. Chem. Soc.* **1974**, *96*, 6211, studied the epr spectrum of the allyl radical at temperatures up to 280° and found no evidence of rotation about a carbon-carbon bond, suggesting that the barrier to rotation is at least 17 kcal/mol.

Note, however, that Shaik, S. S.; Hiberty, P. C.; Lefour, J.-M.; Ohanessian, G. *J. Am. Chem. Soc.* **1987**, *109*, 363, have also questioned the role of electron delocalization in the stabilization of the allyl radical.

Foresman, J. B.; Wong, M. W.; Wiberg, K. B.; Frisch, M. J. *J. Am. Chem. Soc.* **1993**, *115*, 2220, reported that the barrier for rotation about the C-1 - C-2 bond of tetramethylallyl cation and anion, respectively, are 19.4 and 8.6 kcal/mol, respectively, in the gas phase. See also Wiberg, K. B.; Cheeseman, J. R.; Ochterski, J. W.; Frisch, M. J. *J. Am. Chem. Soc.* **1995**, *117*, 6535.

HMO calculation for rotation about allyl radical

a. Input Hückel matrix for starting allyl radical ($H_{2,3}$ = 1.00)

$$\begin{vmatrix} X & 1 & 0 \\ 1 & X & 1 \\ 0 & 1 & X \end{vmatrix} = 0$$

Table of molecular orbitals and energy levels:

	MO 1	MO 2	MO 3
c\E	1.414	.000	-1.414
1	.500	-.707	.500
2	.707	.000	-.707
3	.500	.707	.500

With 2, 1, and 0 electrons in the 3 MOs, E_π = 3 α + 2.828 ß

Values of ρ_i (in bold) and P_{ij} are summarized in the following table:

	1	2	3
1	**1.0000**	.7071	
2	.7071	**1.0000**	.7071
3		.7071	**1.0000**

Free valence indices:

$\mathcal{F}_1$ = 1.025; $\mathcal{F}_2$ = .318; $\mathcal{F}_3$ =1.025

b. Input Hückel matrix for a partial rotation ($H_{2,3}$ = 0.5):

$$\begin{vmatrix} X & 1 & 0 \\ 1 & X & .5 \\ 0 & .5 & X \end{vmatrix} = 0$$

Table of molecular orbitals and energy levels:

	MO 1	MO 2	MO 3
c\E	1.118	.000	-1.118
1	.632	-.447	.632
2	.707	.000	-.707
3	.316	.894	.316

With 2, 1, and 0 electrons in the 3 MOs, E_π = 3 α + 2.236 ß

Values of ρ_i (in bold) and P_{ij} are summarized in the following table:

	1	2	3
1	**1.0000**	.8944	
2	.8944	**1.0000**	.4472
3		.4472	**1.0000**

Free valence indices:

$\mathcal{F}_1$ = .838; $\mathcal{F}_2$ = .614; $\mathcal{F}_3$ =1.508

c. Input Hückel matrix for 90° rotation (transition structure, $H_{2,3} = 0$):

$$\begin{vmatrix} X & 1 & 0 \\ 1 & X & 0 \\ 0 & 0 & X \end{vmatrix} = 0$$

Table of molecular orbitals and energy levels:

	MO 1	MO 2	MO 3
c\E	1.000	.000	-1.000
1	.707	.000	-.707
2	.707	.000	.707
3	.000	1.000	.000

With 2, 1 and 0 electrons in the 3 MOs, $E_\pi = 3\alpha + 2.000\beta$

Values of ρ_i (in bold) and P_{ij} are summarized in the following table:

	1	2	3
1	**1.0000**	1.0000	
2	1.0000	**1.0000**	.0000
3		.0000	**1.0000**

Free valence indices:

$\mathcal{F}_1 = .732$; $\mathcal{F}_2 = .732$; $\mathcal{F}_3 = 1.732$

The activation energy should be 0.828 x 18 kcal/mole = about 15 kcal/mol.

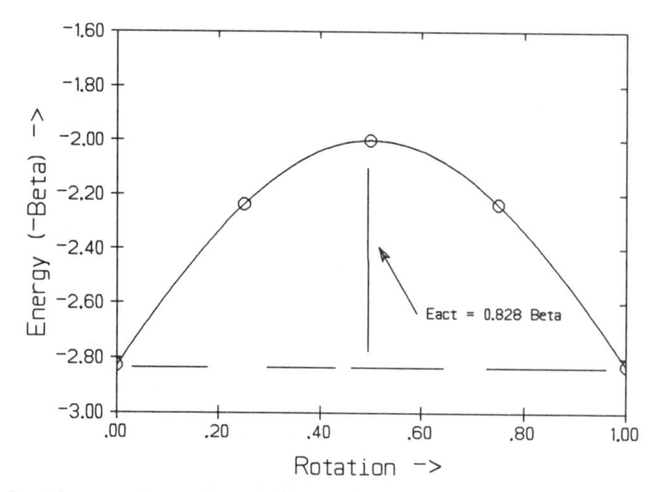

Reaction coordinate diagram for rotation about C1-C2 of allyl radical.

For the benzyl system, the delocalization energy of the benzyl radical is 0.721β. as obtained by the following HMO calculation and substracting the π energy from 7 α + 8 β (6α + 8β for an aromatic ring and α for an electron in a non-interacting p orbital).

Hückel matrix for benzyl radical

$$\begin{vmatrix} X & 1 & 0 & 0 & 0 & 1 & 1 \\ 1 & X & 1 & 0 & 0 & 0 & 0 \\ 0 & 1 & X & 1 & 0 & 0 & 0 \\ 0 & 0 & 1 & X & 1 & 0 & 0 \\ 0 & 0 & 0 & 1 & X & 1 & 0 \\ 1 & 0 & 0 & 0 & 1 & X & 0 \\ 1 & 0 & 0 & 0 & 0 & 0 & X \end{vmatrix} = 0$$

Table of molecular orbitals and energy levels:

c\E	MO 1 2.101	MO 2 1.259	MO 3 1.000	MO 4 .000	MO 5 -1.000	MO 6 -1.259	MO 7 -2.101
1	.500	-.500	.000	.000	.000	.500	.500
2	.406	-.116	-.500	-.378	-.500	-.116	-.406
3	.354	.354	-.500	.000	.500	-.354	.354
4	.337	.562	.000	.378	.000	.562	-.337
5	.354	.354	.500	.000	-.500	-.354	.354
6	.406	-.116	.500	-.378	.500	-.116	-.406
7	.238	-.397	.000	.756	.000	-.397	-.238

With 2, 2, 2, 1, 0, 0, and 0 electrons in the 7 MOs, $E_\pi = 7\,\alpha + 8.721\,\beta$

Values of ρ_i (in bold) and P_{ij} are summarized in the following table:

	1	2	3	4	5	6	7
1	**1.0000**	.5226				.5226	.6350
2	.5226	**1.0000**	.7050				
3		.7050	**1.0000**	.6350			
4			.6350	**1.0000**	.6350		
5				.6350	**1.0000**	.7050	
6	.5226				.7050	**1.0000**	
7	.6350						**1.0000**

Free valence indices:
$\mathcal{F}_1 = .052$; $\mathcal{F}_2 = .504$; $\mathcal{F}_3 = .392$; $\mathcal{F}_4 = .462$; $\mathcal{F}_5 = .392$; $\mathcal{F}_6 = .504$;
$\mathcal{F}_7 = 1.097$

The delocalization energies of the benzyl cation and anion should be the same, so all should have the same rotational barrier: 0.72ß or about 13 kcal/mol.

13. There is no literature reference for this problem.

i. The trends suggest that the $\mathcal{F}_i$ value for the terminal carbon of heptatrienyl should be smaller than that for pentadienyl, while that for octatetraene should be larger than that for hexatriene. The following figure shows the trends in $\mathcal{F}i$ values for additional members of each series, which appear to approach each other asymptotically.

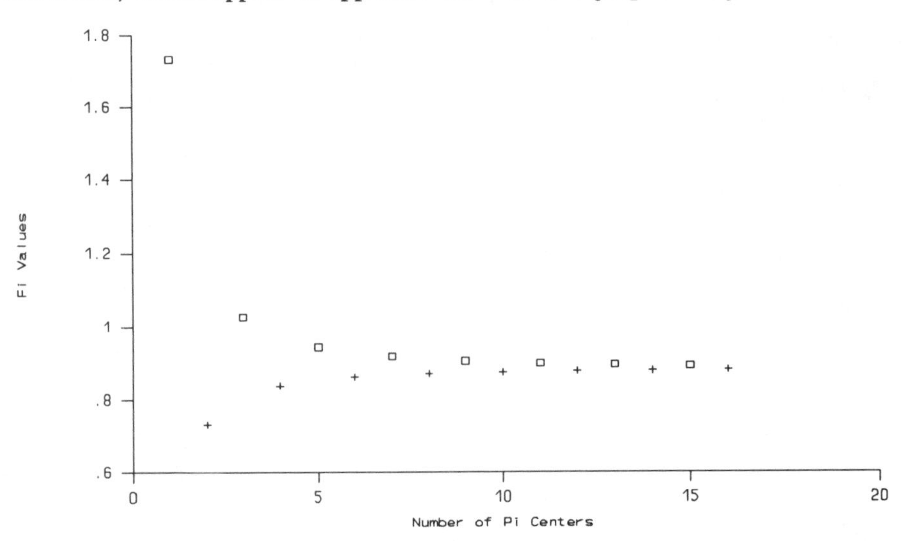

ii. $\mathcal{F}_1$ values are closely related to P_{12} values, which increase as each additional double bond is added to the allyl system. In essence, the $\mathcal{F}_1$ value of a polyenyl system decreases with increasing chain length because the unpaired electron density is distributed over a larger molecule, meaning that the overlap of the two terminal π centers is more like that in a polyene. In polyenes $\mathcal{F}_1$ values decrease with increasing chain length because P_{12} values decrease (and P_{23} values increase) with increasing chain length.

iii. The large value of $\mathcal{F}$ for the methylene carbon atoms of 1,2-dimethylene-3,5-cyclohexadiene can be rationalized by drawing a resonance structure with an aromatic ring and two —$CH_2\cdot$ centers. This model suggests that the two methylene carbon atoms should have considerable radical character, hence large $\mathcal{F}$ values.

iv. If COT is tub-shaped, the double bonds should be essentially noninteracting, and the $\mathcal{F}_i$ values should be the same as those in ethene.

14. For phenanthrene:

$$RE = 2(5\Gamma_1 + 2\Gamma_2)/5 = 2.32\ \Gamma_1 = 1.944 \text{ eV}.$$

This compares with 1.933 by SCF-MO and a value of 5.448 ß or 4.25 eV from HMO if ß is assumed to be 18 kcal/mol = 0.78 eV. See HMO calculation for phenanthrene above.

For styrene:

$$RE = 2(\Gamma_1 + 0\Gamma_2)/2 = \Gamma_1 = 0.838 \text{ eV}$$

For pyrene:

$$RE = 2(6\Gamma_1 + 4\Gamma_2)/6 = 2.12 \text{ eV}$$

(Note: denote rings as A, B, C, and D. Deleting A leaves a polyene with SC = 1; ditto for D by symmetry. Deleting B gives 1,3-divinylbenzene with SC = 2; ditto for C. Now delete A and B at the same time to give hexatriene: SC_{AB} = 1; the same will be true for SC_{CD}. SC_{AC} = SC_{BD} = 1.) The Resonance energy for pyrene can also be calculated from the equation

$$RE\ (ev) = 1.185 \ln (CSC[R]) = 1.185 \ln 6 = 2.12 \text{ eV}$$

see Herndon, W. C. *J. Am. Chem. Soc.* **1976**, *98*, 887 and references therein. Herndon has developed a number of applications for the method. One can calculate the ionization potentials of π-molecular hydrocarbons (Herndon, W. C. *J. Am. Chem. Soc.* **1976**, *98*, 887) as well as bond orders and bond lengths (Herndon, W. C. *J. Am. Chem. Soc.* **1974**, *96*, 7605).

15. i. The cycloheptatrienylium (tropylium) ion was reported by Doering, W. v. E.; Knox, L. H. *J. Am. Chem. Soc.* **1954**, *76*, 3203. Both chemical and spectroscopic data suggest that the ion is aromatic.

ii. The slow reaction of 5-iodocyclopentadiene with silver ion is consistent with antiaromatic character for the cyclopentadienyl cation. Breslow, R.; Hoffman, Jr., J. M. *J. Am. Chem. Soc.* **1972**, *94*, 2110.

iii. The cyclooctatetraenyl dianion was reported by Katz, T. J. *J. Am. Chem. Soc.* **1960**, *82*, 3784, 3785. The NMR spectrum suggests aromatic character.

16. As the overlap decreases along two sides (leaving the other sides fixed at $\beta = 1$), the total energy stays the same. As *one* side is stretched, the π energy approaches that of 1,3-butadiene, so the HMO delocalization energy for cyclobutadiene should be greater for a trapezoidal structure than for a rectangular structure. Of course, this approach ignores molecular symmetry and the restrictions of a σ framework.

17. The experimental heat of formation of glyoxal is -50.66 kcal/mol (Cox, J. D.; Pilcher, G. *Thermochemistry of Organic and Organometallic Compounds*; Academic Press: New York, 1970; p. 223 and references therein to unpublished work).
The calculation is: $7.435 \times 2 - 0.604 \times 2 - 32.175 \times 2 - 29.38 \times 0 = -50.69$

18. Baird, N. C. *J. Chem. Educ.* **1971**, *48*, 509.

a. ΔDRE is smaller for addition across the 9,10 position of anthracene, so this pathway is favored. This problem is similar to the comparison of addition across the 1,4 and 9,10 positions of anthracene discussed in Chapter 4.

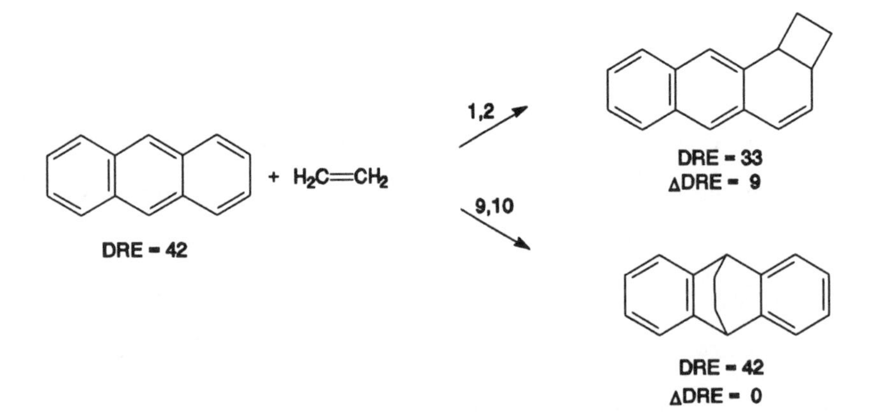

b. This is similar to the addition of H_2 across the same two positions of phenanthrene as discussed by Baird on page 512 of the paper cited above. For the addition across the 9,10 positions, the DRE of the product is taken to be that of 2 benzenes, which is 42 kcal/mol. Since the DRE of phenanthrene is 49, then ΔDRE = 7. For addition across the 1,2-positions, the product has DRE = 33, since that is the DRE of naphthalene. For this reaction, ΔDRE = 16, so this reaction is uphill by 16 kcal/mol. Thus addition across 9,10 position is favored.

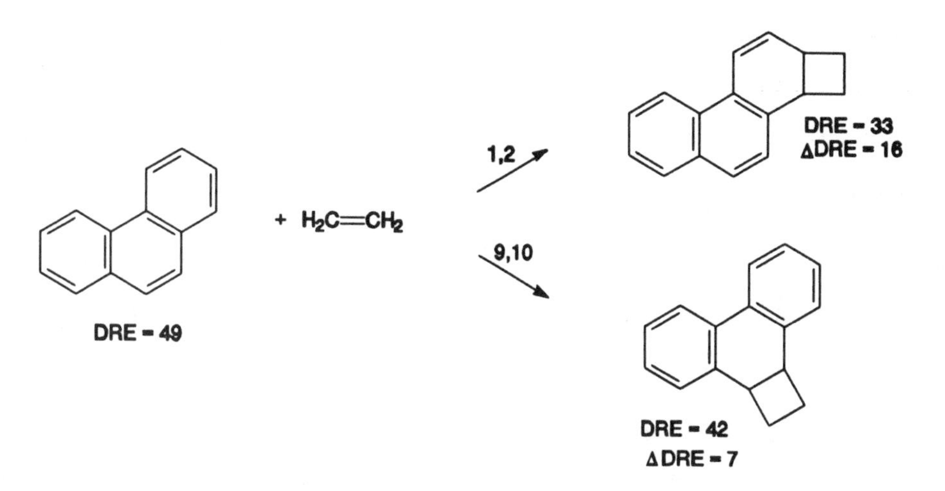

19. Iyoda, M.; Kurata, H.; Oda, M.; Okubo, C.; Nishimoto, K. *Angew. Chem., Int. Ed. Engl.* **1993**, *32*, 89.

Two electrons can be added to give a species with a net charge of 2- but which has an aromatic cyclopropenyl cation center ring in which each carbon is bonded to a carbon that is part of an aromatic cyclopentadienyl anion ring. The stability of the dianion results from the stabilization due to the aromatic rings.

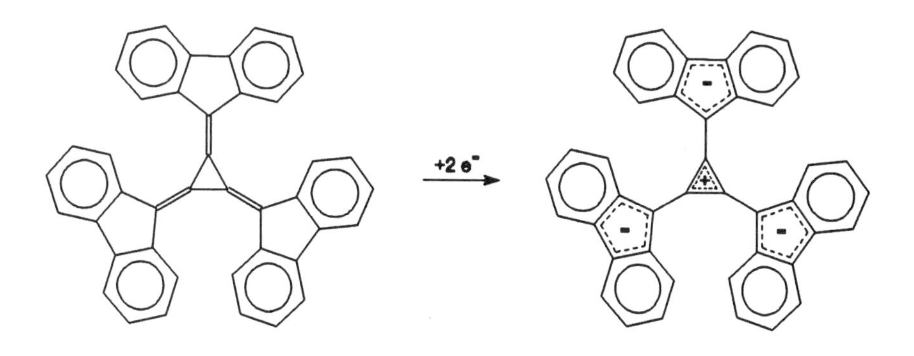

20. Binsch, G. *Top. Stereochem.* **1968**, *3*, 97 (see especially pp. 134-146).

Shift of electron density from the three-membered ring to the five-membered ring (as shown below) results in a relatively stable structure with two aromatic rings. Therefore, there is considerable single bond character to the bond connecting the two rings, and the barrier to rotation is lowered in comparision with normal double bonds.

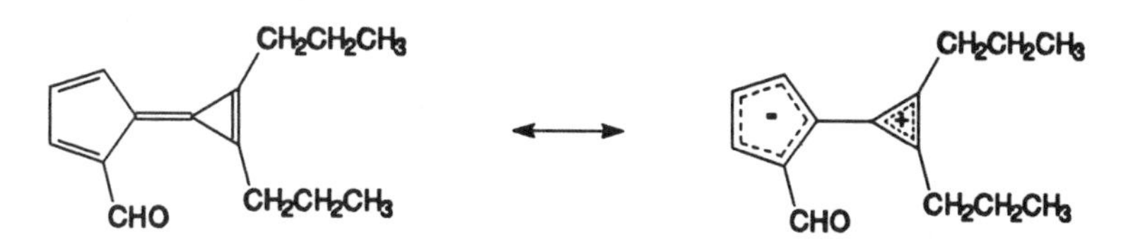

21. Sekiguchi, A.; Ebata, K.; Kabuto, C.; Sakurai, H. *J. Am. Chem. Soc.* **1991,** *113*, 7081.
    The compound is antiaromatic.

22. Baird, N. C. *J. Chem. Educ.* **1971,** *48*, 509. (See especially p. 511.)
    No.

23. There is no single literature reference for this problem.
    Benzene is aromatic because we define it to be so. That definition is based both on its physical properties (so in that sense benzene is aromatic because it is stable) and on our theoretical analysis of its structure and bonding (so in that sense benzene is stable because it is aromatic.) Therefore, there is not a simple answer to this question.

24. There is no literature reference for this problem.
    Aromaticity is a useful model for a variety of reasons. It also serves as a concrete example of a paradigm — a solution to a particular piece of the puzzle that gives us the form for solution of other pieces of the puzzle. Yet because our paradigms and models are limited, it is not a complete solution to all related problems.

Chapter 5

1. Wentrup, C. *Reactive Molecules*; John Wiley & Sons: New York, 1984; p. 41.
   C-C bonds use more *p* character to avoid ring strain, therefore C-H bonds use more *s* character. Planar cyclopropyl radical would be *sp*2, so rings would use C-C bonds with more *s* character and thus the ring would have more ring strain.

2. Egger, K. W.; Cocks, A. T. *Helv. Chim. Acta* **1973**, *56*, 1537, particularly p. 1539. Radicals do not tend to rearrange, so the experimental data are probably correct. The carbocations tend to rearrange quite rapidly, so the values are more uncertain. Also, the carbocations are more susceptible to substituent effects and other structural effects.

3. For a discussion, see Wentrup, C. *Reactive Molecules*; John Wiley & Sons: New York, 1984; pp. 33-34, which cites the original literature for these compounds.
   Structure **57** has a stable (singlet) resonance structure that has no radical character, so it exhibits predominantly singlet character. Structure **58** cannot generate a Kekulé resonance structure (that is, a structure having only double and single bonds) that does not have radical character, so it is a triplet species.

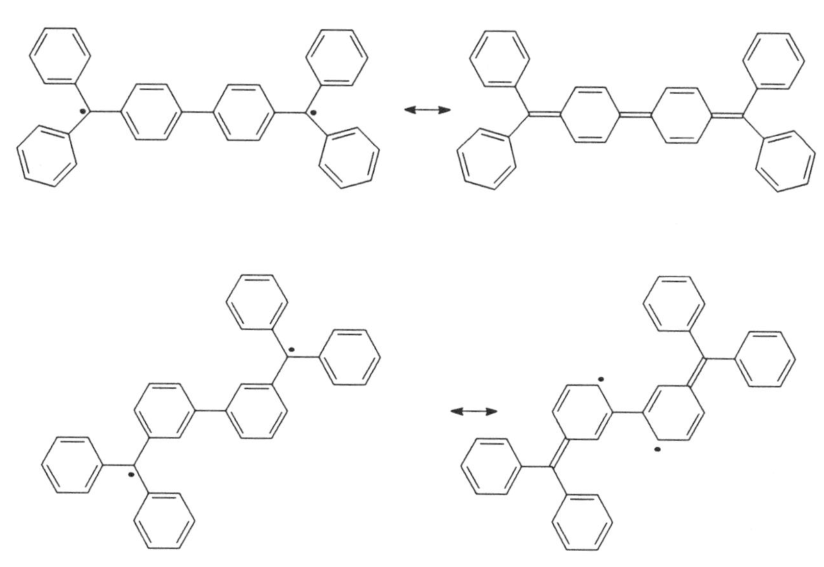

4. Ayscough, P. B. *Electron Spin Resonance in Chemistry*; Methuen & Co,: London, 1967; p. 298; Bunce, N. J. *J. Chem. Educ.* **1987**, *64*, 907 (especially p. 910).
   The electron paramagnetic resonance spectrum of the benzyl radical has been interpreted to mean that about 50% of the unpaired electron density is on the benzylic carbon atom, while 15.8% is on each of the two *ortho* carbon atoms and 18.6% is on the *para* carbon atom. See the discussion by Fleming, I. *Frontier Orbitals and Organic Chemical Reactions*; Wiley-Interscience: London, 1976; p. 60.

5. Walling, C.; Cooley, J. H.; Ponaras, A. A.; Racah, E. J. *J. Am. Chem. Soc.* **1966**, *88*, 5361.
78% methylcyclopentane, 7% 1-hexene and a trace of cyclohexane.

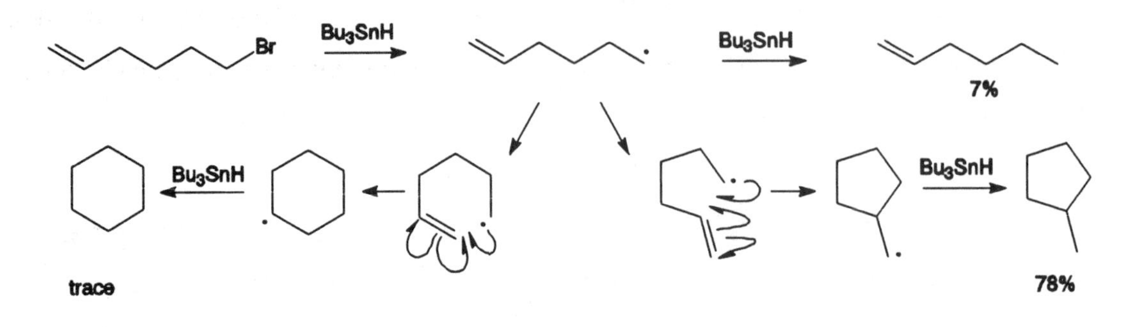

6. Olah, G. A. *Carbocations and Electrophilic Reactions*; John Wiley & Sons: New York, 1974; p. 20.
Using the formula $J_{^{13}C\text{-}H} = 500/(1 + \lambda^2) = 169$ Hz, $1 + \lambda^2 = 2.958$.
Then percent *s* character $= 100\%/(1 + \lambda^2) = 33.8\%$.
Therefore, the carbon orbital is slightly more *s*-like than an sp^2 hybrid orbital.

7. Olah, G. A. *Carbocations and Electrophilic Reactions*; John Wiley & Sons: New York, 1974; p. 63; Olah, G. A.; White, A. M. *J. Am. Chem. Soc.* **1967**, *89*, 3591.

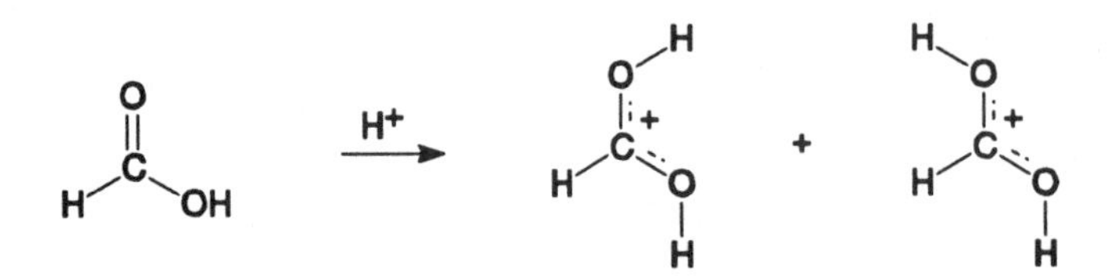

8. Olah, G. A.; Mateescu, G. D.; Wilson, L. A.; Gross, M. H. *J. Am. Chem. Soc.* **1970**, *92*, 7231.
Resonance stabilization delocalizes the charge.

9. Olah, G. A.; Prakash, G. K. S.; Williams, R. E.; Field, L. D.; Wade, K. *Hypercarbon Chemistry*; John Wiley & Sons: New York, 1987; p. 153 and reference therein to unpublished results of M. Saunders and coworkers.
Rapid proton shift from one end of the molecule to the other makes the two sets of methyl protons equivalent.

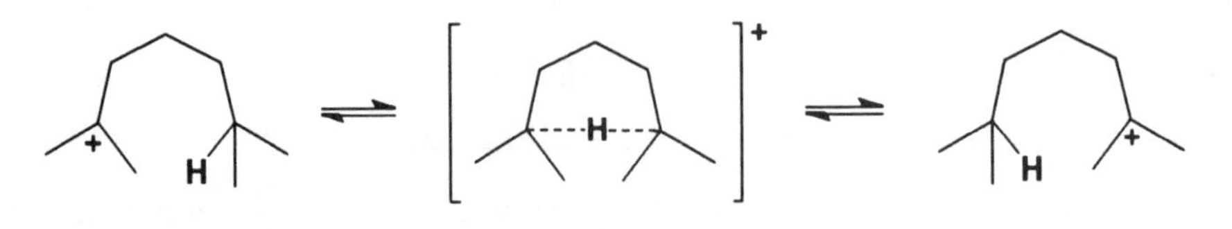

10. Olah, G. A.; Prakash, G. K. S.; Sommer, J. *Superacids*; Wiley-Interscience: New York, 1985; p. 84. The authors suggest the isomerization involves a primary carbocation, then a protonated cyclopropane, then a linear secondary carbocation. The latter can go back to a *t*-butyl cation with a change in the position of the carbon atoms. Similar rearrangements are discussed by Saunders, M.; Vogel, P.; Hagen, E. L.; Rosenfeld, J. *Acc. Chem. Res.* **1973,** *6,* 53.

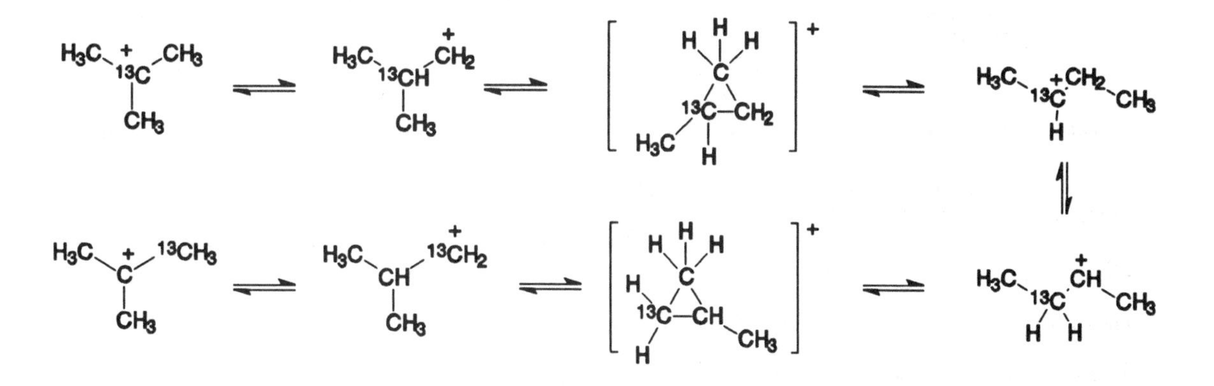

11. Walborsky, H. M.; Periasamy, M. P. *J. Am. Chem. Soc.* **1974,** *96,* 3711.

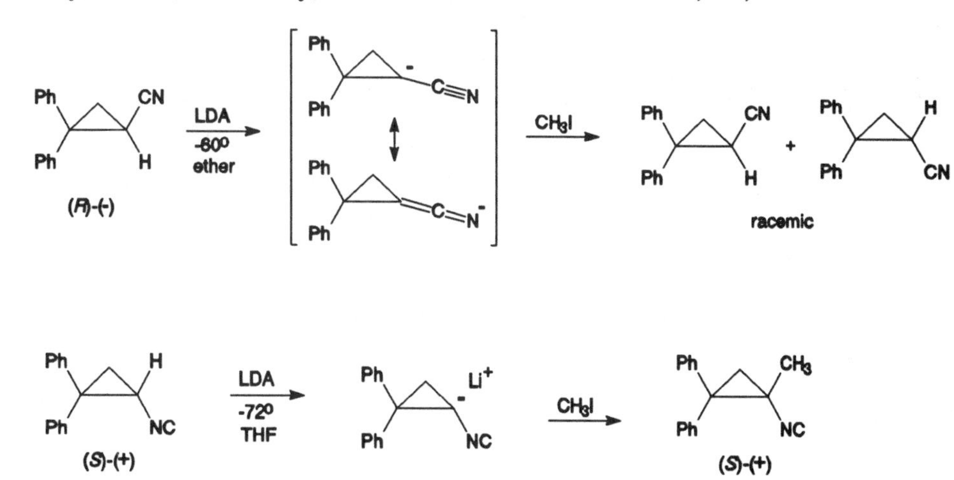

Ph H Ph NC (S)-(+) LDA -72° THF Ph Li+ Ph NC CH3I Ph CH3 Ph NC (S)-(+)

The cyano group is able to stabilize and flatten the carbanion by resonance. However, the iso-cyano group cannot form a double bond with the ring, because to do so would involve resonance of the type

$$\overset{-}{C}-\overset{+}{N}\equiv\overset{-}{C}: \longleftrightarrow C=\overset{+}{N}=\ddot{C}:^{2-}$$

12. Zimmerman, H. E.; Zweig, A. *J. Am. Chem. Soc.* **1961**, *83*, 1196. (see especially p. 1198) Intramolecular reaction converts the alkyllithium to an aryllithium, which then reacts with CO_2 to give (after protonation) the benzoic acid derivative.

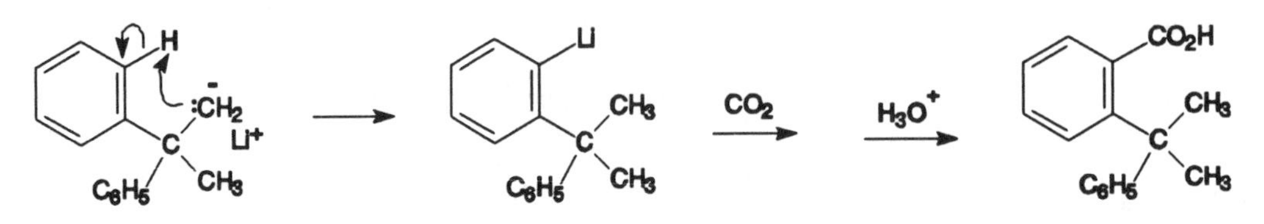

13. Crawford, R. J.; Erman, W. F.; Broaddus, C. D. *J. Am. Chem. Soc.* **1972**, *94*, 4298; also see Bates, R. B.; Ogle, C. A. *Carbanion Chemistry*; Springer-Verlag: Berlin, 1983; p. 24. The product reflects the greater stability in solution of the carbanion bearing fewer alkyl substituents.

14. Fitjer, L.; Quabeck, U. *Angew. Chem., Int. Ed. Engl.* **1987**, *26*, 1023.

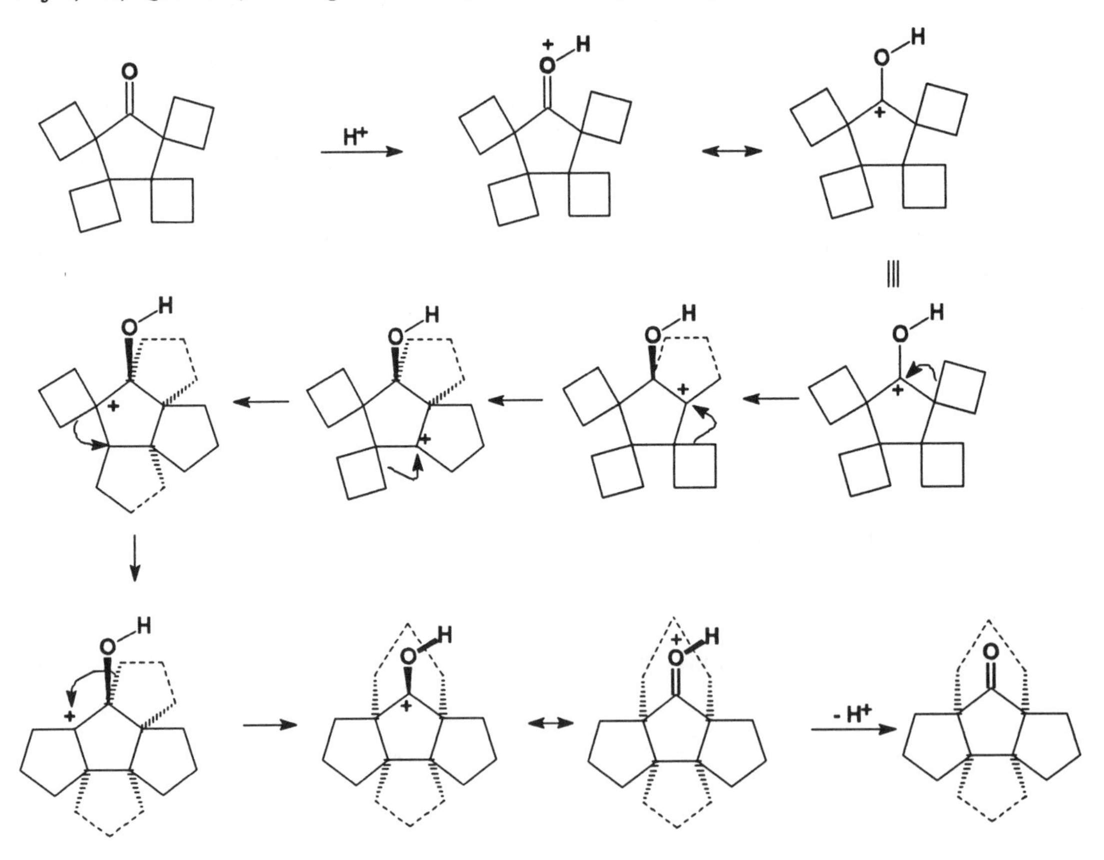

15. Oldroyd, D. M.; Fisher, G. S.; Goldblatt, L. A. *J. Am. Chem. Soc.* **1950**, *72*, 2407. Also see the discussion by Walling, C. in *Molecular Rearrangements*, Part I; de Mayo, P., Ed.; Wiley-Interscience: New York, 1963; pp. 407-455 (particularly p. 440).

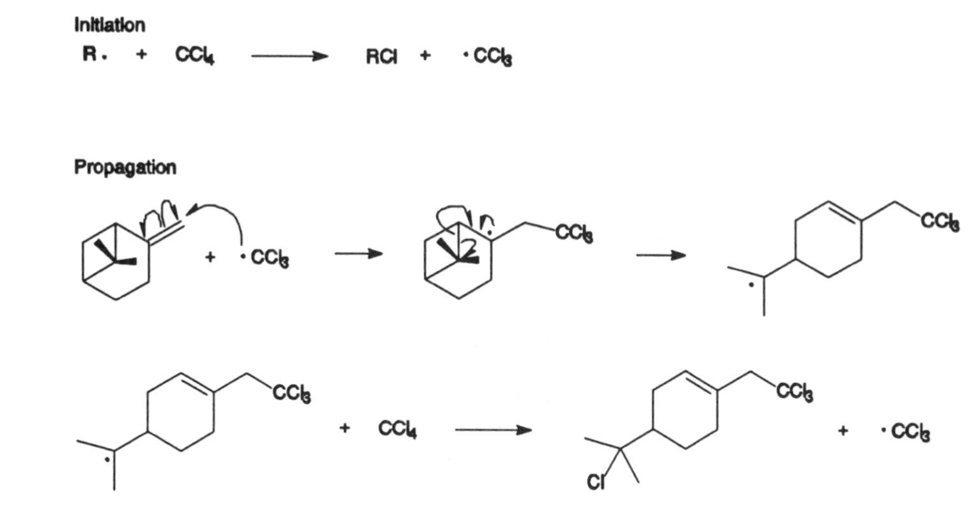

16. For a discussion, see Moss, R. A.; Jones, Jr., M. in *Reactive Intermediates*, Vol. 1; Jones, Jr., M.; Moss, R. A., eds.; Wiley-Interscience: New York, 1978; pp. 67 - 116 (especially p. 97 and references therein); Wentrup, C. *Reactive Molecules*; John Wiley & Sons: New York, 1984; p. 238 and references therein; Hoffmann, R. W.; Reiffen, M. *Chem. Ber.* **1976**, *109*, 2565. Electron delocalization from the oxygens to the carbenic p orbital means that the carbene is not as electrophilic and is less likely to react with nucleophiles.

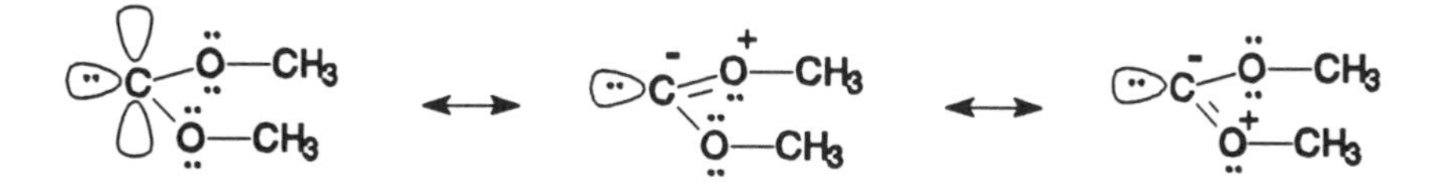

17. Perkins, M. J. *Adv. Phys. Org. Chem.* **1980**, *17*, 1.

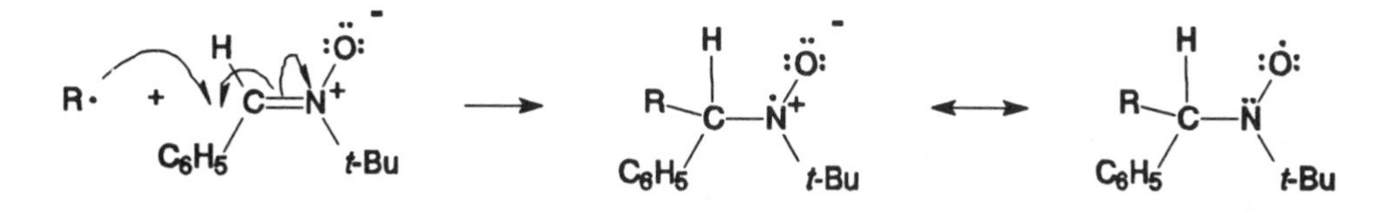

18. Friedman, L.; Shechter, H. *J. Am. Chem. Soc.* **1960**, *82*, 1002.

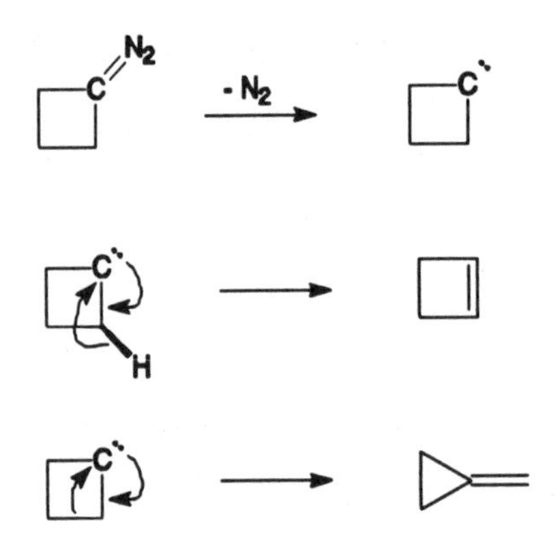

19. Friedman, L.; Shechter, H. *J. Am. Chem. Soc.* **1961**, *83*, 3159.
The first two products are formed by insertion of the carbene into a transannular carbon-hydrogen bond. The last two products are formed by hydrogen shift.

20. Jones, R. R.; Bergman, R. G. *J. Am. Chem. Soc.* **1972**, *94*, 660. The cyclization of a 1,5-hexadiyn-3-ene to a 1,4-dehydrobenzene is known as the Bergman cyclization.

a. Five other structures are possible. Four of them (as well as **62**) have both a C_2 axis and a plane of symmetry. However structure (v) below has only a C_2 axis.

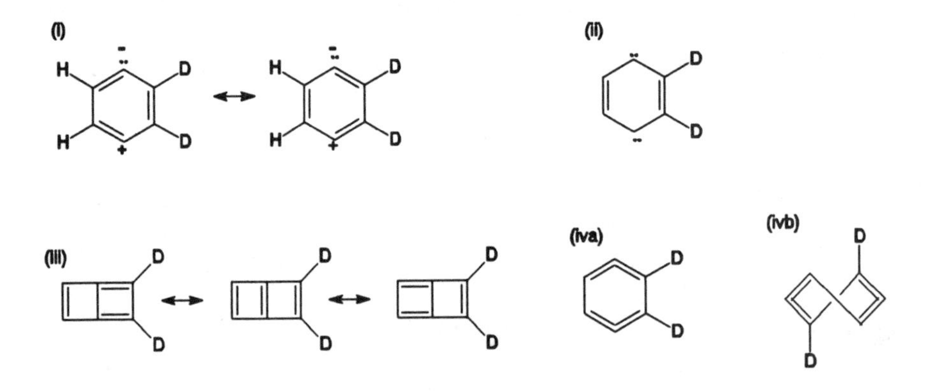

b. The formation of benzene is most readily explained as being the product of radical atom abstraction and coupling reactions. Intermediates with ionic character would be expected to react with methanol to yield anisole, and those with allenic character would not readily abstract hydrogen atoms from alkane solvent molecules.

21. There is no literature reference for this problem.
Inconsistencies between expected and observed experimental results usually prompt investigators to examine the validity of their experimental data, the design of their experiments, and then the assumptions on which the experiments are based. It seems unnecessary to routinely repeat prior experiments unless such inconsistencies suggest a likelihood that prior work was in error.

Chapter 6.

1. The recombination of the fragments of dissociation of the molozonide might be occurring within a solvent cage. Increasing the concentration of the reactant should increase the chance of finding cross ozonolysis products.

   See Murray, R. W. *Acc. Chem. Res.*, **1968**, *1*, 313 and references therein. See also Murray, R. W.; Story, P. R.; Loan, Jr., L. D. *J. Am. Chem. Soc.* **1965**, *87*, 3025.

2. Long, F. A.; Pritchard, J. G. *J. Am. Chem. Soc.* **1956**, *78*, 2663.

   Both reactions are highly but not completely regioselective. In each case most of the labeled oxygen is found on the carbon atom bearing the two hydrogen atoms, although some product is formed in which the labeled oxygen is bonded to the carbon atom bearing the methyl group(s). These results are consistent with a mechanism in which hydroxide ion attacks the epoxide preferentially at the less highly substituted carbon atom by an S_N2 pathway.

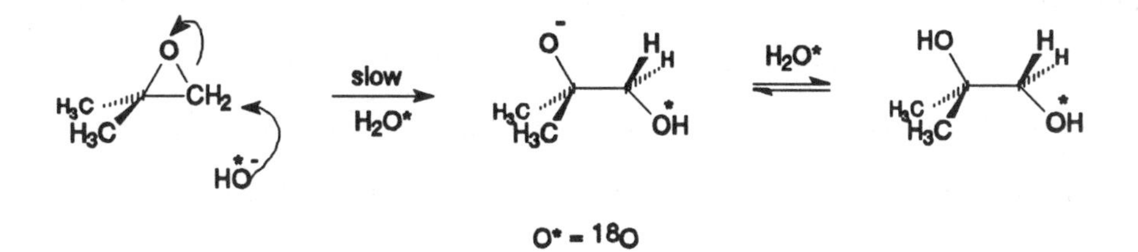

3. There is no literature reference for this problem.

   Set $d[C]/dt = k_1[A][B] - k_2[C] - k_{-1}[C] = 0$, so that $[C] = k_1[A][B]/(k_2 + k_{-1})$, then let $(k_2 + k_{-1}) \approx k_{-1}$.

4. Bartlett, P. D.; Trachtenberg, E. N. *J. Am. Chem. Soc.* **1958**, *80*, 5808.

   For compound A, $\Delta H^\ddagger$ is 10.2 kcal/mol, while $\Delta S^\ddagger$ is -30 e.u. For compound B, the two values are 31.3 kcal/mol and +25.4 e.u. Thus, a lower temperature favors the reactant with the smaller activation enthalpy, while the higher temperature favors the reactant for which the activation entropy is more positive.

5. The data are from Frost, A. A.; Pearson, R. G. *Kinetics and Mechanism*, 2nd ed.; John Wiley & Sons, Inc.: New York, 1961; pp. 101 and 105. For a discussion of enthalpy and entropy terms in the Diels-Alder reaction, see Chung, Y.-S.; Duerr, B. F.; Nanjappan, P.; Czarnik, A. W. *J. Org. Chem.* **1988**, *53*, 1334.

   Apparently the entropy of the transition structure is very nearly the same as that of the dimer. Therefore, $\Delta S^\ddagger$ for fragmentation is zero. The entropy of the dimer is much less than that of two dicyclopentadiene molecules, however, so $\Delta S^\ddagger$ for dimerization is quite negative.

6. LeFevre, G. N.; Crawford, R. J. *J. Org. Chem.* **1986**, *51*, 747.

The authors stated that "It is interesting to note that in those cases wherein the pivoting carbon is stabilized by resonance, e.g., a vinyl or phenyl group, a negative entropy of activation is observed. This is to be expected, since a loss of rotational freedom is required for the resonance interaction." Thus, the mechanism is presented as a diradical and not a concerted mechanism. The vinyl substituents stabilize the radical intermediate, thus lowering the activation enthalpy. The earlier transition state reduces the activation entropy. For the latter set of compounds, the requirement that the freely-rotating vinyl group become planar with the radical center in the transition state means that the activation entropy is negative.

7. Kluger, R.; Brandl, M. *J. Org. Chem.* **1986**, *51*, 3964.

Plot log k/T versus $1/T$; The slope is -4,871.6, so $\Delta H^{\ddagger} = -4871.6 \times -4.576/1000 = 22.3$ kcal/mol. $\Delta S^{\ddagger} = R \times 2.303 \times \log(k/T) + \Delta H^{\ddagger}/T - R \times 2.303 \times \log(\kappa k/h) = -4.05$ e.u. at 25°. Alternatively, plot ln k vs. $1/T$ to calculate E_a, which is found to be 22.9 kcal/mol. Then $\Delta H^{\ddagger}$ is estimated from the formula $\Delta H^{\ddagger} = E_a - RT = 22.3$ kcal/mol at 25°. Then $\Delta S^{\ddagger} = 4.576 \log A - 60.53 = -3.9$ e.u. Since $\Delta S^{\ddagger}$ is negative, the mechanism is probably concerted. A possible mechanism is shown below:

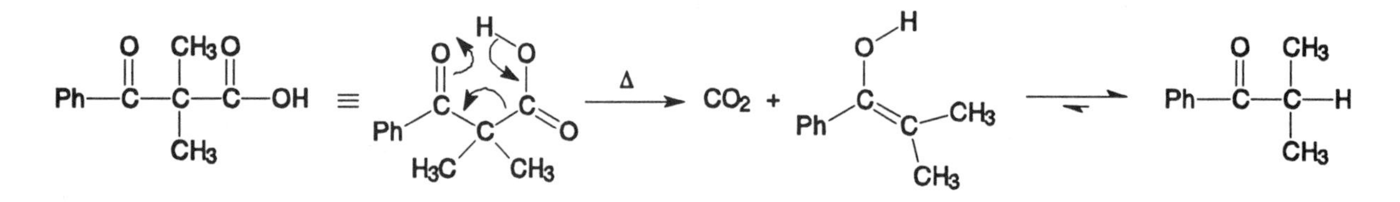

8. Gano, J. E.; Lenoir, D.; Park, B.-S.; Roesner, R. A. *J. Org. Chem.* **1987**, *52*, 5636.

The literature value is $E_a = 40.4 \pm 1.7$ kcal/mol. (An E_a value of 42.6 kcal/mol is calculated with the subset of experimental data cited in the problem.) The E_a is about 22 kcal/mol less than that for 2-butene, which means that the *cis-t*-butyl groups lower the barrier for rotation about the double bond. However, the authors conclude that at least 9 kcal/mol of strain is still present in the transition structure for the rotation.

9. Wiberg, K. B.; Caringi, J. J.; Matturro, M. G. *J. Am. Chem. Soc.* **1990**, *112*, 5854.

The literature values are $\Delta H^{\ddagger} = 45.0 \pm 1.1$ kcal/mol, $\Delta S^{\ddagger} = 2.2$ e.u., $E_a = 46.3 \pm 1.3$ kcal/mol, and log $A = 14.0$.

10. Bartlett, P. D.; Wu, C. *J. Org. Chem.* **1985**, *50*, 4087.

The data suggest that the entropy is decreased in the transition structure, so a nondissociative mechanism is favored.

11. a) selectivity is greater because the transition state lies further to right with Br; therefore there is more radical character developed, so differences in radical stability are more important in determining differences in activation energies;

    b) k_H/k_D is greater for bromination because the transition structure lies further to right, so a more symmetrical transition structure is developed;

    c) Wiberg, K. B.; Slaugh, L. H. *J. Am. Chem. Soc.* **1958**, *80*, 3033. The values are 4.86, 2.67, and 1.81, respectively.

12. Jones, J. M.; Bender, M. L. *J. Am. Chem. Soc.* **1960**, *82*, 6322.

    This is a ß secondary effect. For the dissociation, K_H/K_D = 1.44. "At each concentration, the deuterium compound is less dissociated than the hydrogen compound."

13. Pocker, Y. *Proc. Chem. Soc.* **1960**, 17.

    This should be an inverse 2° isotope effect, so k_H/k_D should be less than 1. The literature value is 0.98.

14. In ethanol solvent stabilization of the carboxylate ion is less effective than is the case in water, so the effect of a substituent on the benzene ring is more signficant.

15. There is no literature reference for this problem.

    Rewrite equation so that log k_0 is a constant.

16. Overman, L. E.; Petty, S. T. *J. Org. Chem.* **1975**, *40*, 2779.

    The value calculated for ρ is 1.76. The data indicate that "negative charge is developed on both sulfur atoms in the transition state as bond making is somewhat advanced over bond breaking."

17. Dietze, P. E.; Underwood, G. R. *J. Org. Chem.* **1984**, *49*, 2492.

    a) and b) Both rate constants and pK_a values correlate better with σ^-. Note: plot $-pK_a$ values, not pK_a values. The investigators report that there is a good correlation of rate constants with σ^-, with ρ^- = -2.54, suggesting that the phenoxide undergoes significant change in charge distribution. Therefore the N-Cl bond is thought to be substantially broken in the transition state, consistent with nucleophilic attack on the Cl.

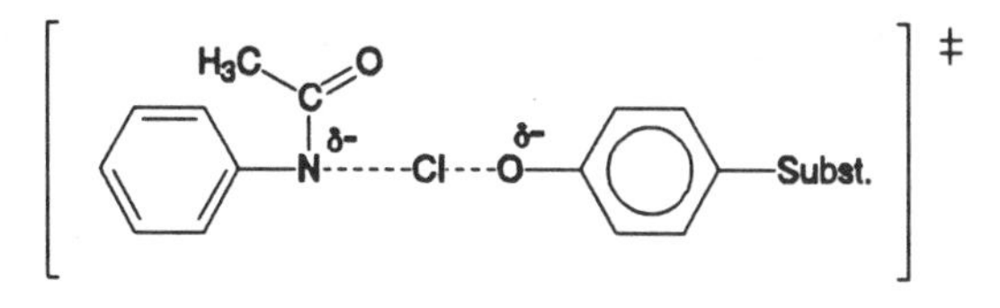

The value of ρ for ionization is found to be 2.8 based on the pK_a values measured under the reaction conditions.

18. Whitworth, A. J.; Ayoub, R.; Rousseau, Y.; Fliszár, S. *J. Am. Chem. Soc.* **1969**, *91*, 7128.
At any one temperature, log (k/k_0) = $(-0.91 \pm 0.03) \times \sigma$. The addition is strongly electrophilic. That is, the reaction is facilitated by electron-donating substituents on the carbon-carbon double bond.

19. Choe, J.-I.; Srinivasan, M.; Kuczkowski, R. L. *J. Am. Chem. Soc.* **1983**, *105*, 4703.
There is an inverse 2° isotope effect, consistent with the conversion of sp^2 to sp^3 hybridization at the transition state. Since both values are the same, the data suggest nearly the same extent of bonding to both of the carbon atoms in the transition structure.

20. Hill, R. K.; Conley, R. T.; Chortyk, O. T. *J. Am. Chem. Soc.* **1965**, *87*, 5646.

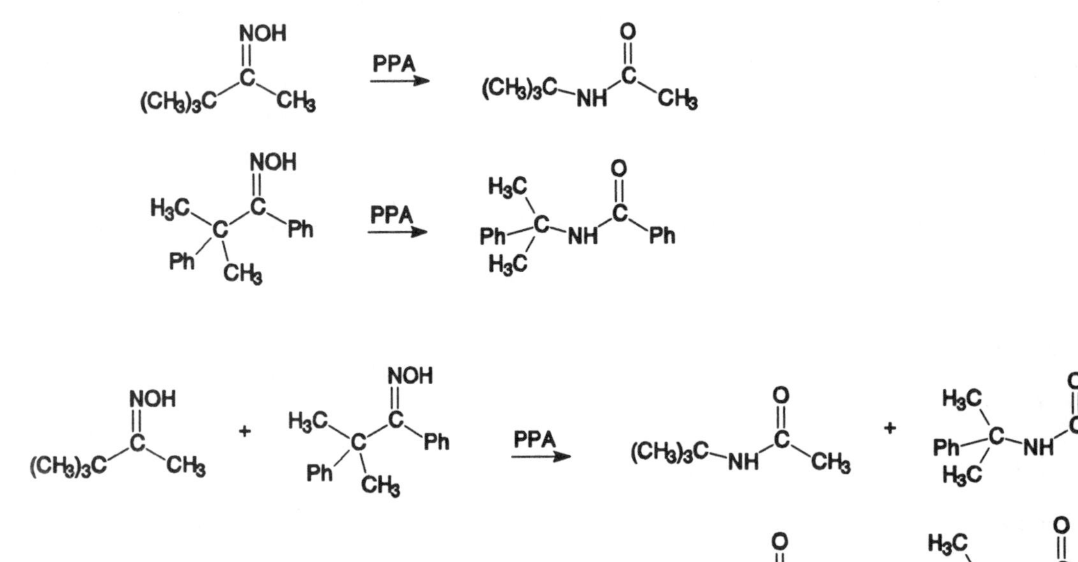

a. The observation of cross products strongly suggests that at least some product formation results from a dissociative pathway.

b. Optically active oximes that could produce achiral carbocations upon dissociation were investigated in the same study. Observation of racemic Beckmann rearrangement product is consistent with a dissociative pathway.

c. The dissociative pathway is most likely for those oximes with structural features that facilitate carbocation formation.

21. Berliner, E.; Altschul, L. H. *J. Am. Chem. Soc.* **1952,** *74,* 4110. See also Leffler, J. E.; Grunwald, E. *Rates and Equilibria of Organic Reactions*; John Wiley and Sons, Inc.: New York, 1963; pp. 324 - 342.

For each compound, the experimental data are used to calculate $\Delta G^{\ddagger} = E_a - RT - T\Delta S^{\ddagger}$ for a series of temperatures. Graphing the values of $\Delta G^{\ddagger}$ vs. T leads to a determination that the lines for all of the compounds cross around 750 K.

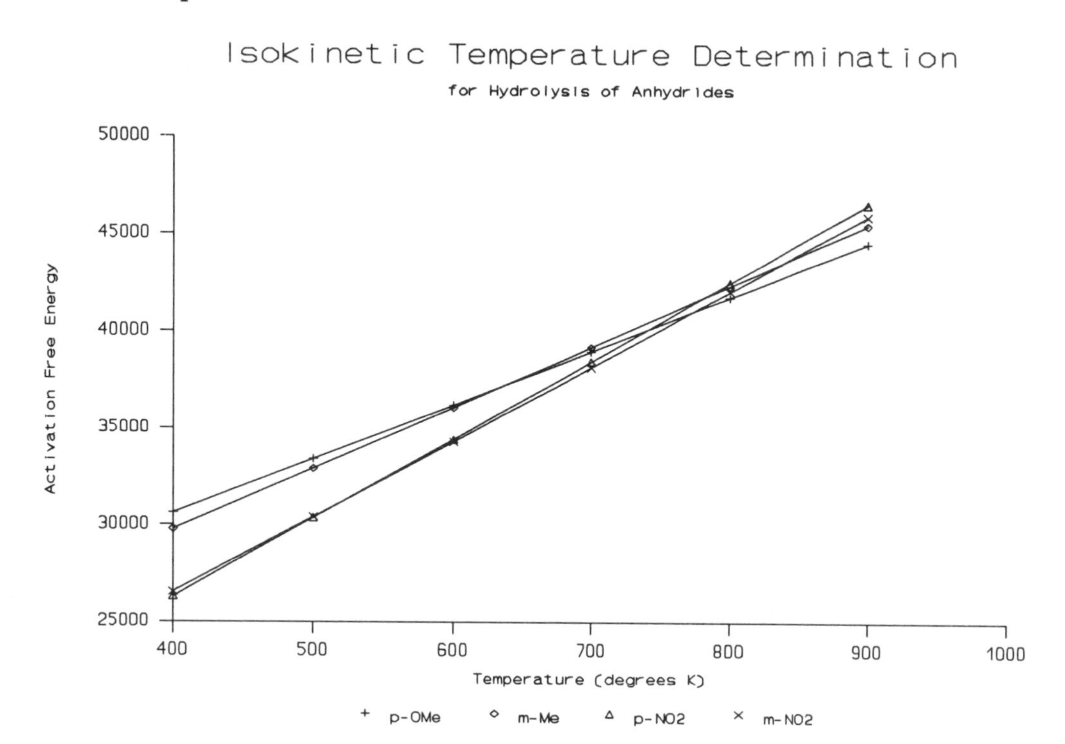

Chapter 7

1. Barlin, G. B.; Perrin, D. D. *Quart. Rev. Chem. Soc.* **1966**, *20*, 75.

$$pK_a = 4.20 - \Sigma\sigma = 4.20 - (-.37 + .115) = 4.46$$

2. Taft, R. W.; Bordwell, F. G. *Acc. Chem. Res.* **1988**, *21*, 463.

$$\Delta G° = -RT \ln K_a = 2.303\ RT\,(-\log K_a) = 2.303\ RT\,pK_a$$

$$\text{Therefore, } pK_a = \Delta G° \text{ (in kcal/mol)}/1.364.$$

Using the data from Table 7.4, the pK_a values of acetic and propionic acid in the gas phase are found to be 250.4 and 249.6, respectively. This is a difference of 0.8 pK units, with propionic acid being more acidic. The data in Table 7.1 indicate that the pK_a values of acetic and propionic acid in aqueous solution are 4.76 and 4.87, respectively. This is a difference of 0.11 pK units, with acetic acid being more acidic. The magnitude of the pK_a values is smaller in solution because the solvent helps stabilize the ion (resulting in much smaller pK_a values), so the effect of the substituent is less important. Furthermore, propionic acid is more acidic in the gas phase because the replacement of a ß-hydrogen with a methyl group provides greater dispersal of the negative charge in the gas phase. In solution, however, the methyl substituent causes greater disruption of the solvent shell around the carboxylate ion, reducing the acidity of the larger acid.

3. The first substitution (from methanol to ethanol) reduces the calculated $\Delta G°$ value by 3.01 kcal/mol; the second (from ethanol to isopropyl alcohol) by 1.97 kcal/mol, and the third (from isopropyl alcohol to *t*-butyl alcohol) by 1.23 kcal/mol. Each successive substitution has less effect than the previous one because the effect of a methyl group in delocalizing the charge is less significant in a large alkyl group (where there is already signficant charge delocalization) than in a smaller group where the charge is not so extensively delocalized.

4. Wheeler, O. H. *J. Am. Chem. Soc.* **1957**, *79*, 4191.

   Cyclobutanone, K_d = 1.11. Angle strain is relieved by conversion of the sp^2 (120° bond angle preferred) carbonyl carbon to sp^3 (109.5° bond angle preferred) hybridization.

   Cyclopentanone, K_d = 15.1. Angle strain is not appreciably relieved by hemiacetal formation, and some steric hindrance between the methoxy group and the ring hydrogen atoms results from hemiacetal formation.

   Cyclohexanone, K_d = 2.1. Some angle strain is reduced by hemiacetal formation, but some intramolecular van der Waals repulsion is introduced.

5. Use the relationship in equation 7.15 and substitute $-pK_b = \log K_b = pK_{BH^+} - 14$.

6. Wiseman, J. S.; Abeles, R. H. *Biochem.* **1979**, *18*, 427.

   a. Hydrolysis can occur by the mechanisms in Figures 7.24 and 7.25.

   b. Cyclopropane exists almost exclusively in the form of the hydrate because of the the relief of angle strain upon conversion of the sp^2- to sp^3-hybridized carbonyl carbon atom in the cyclopropanone ring.

7. Cardwell, H. M. E.; Kilner, A. E. H. *J. Chem. Soc.* **1951**, 2430 and references therein.

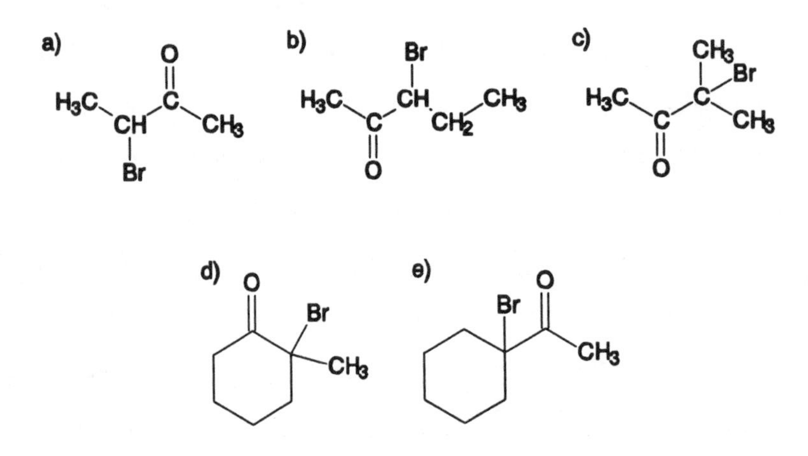

8. Levine, R.; Hauser, C. R. *J. Am. Chem. Soc.* **1944**, *66*, 1768.

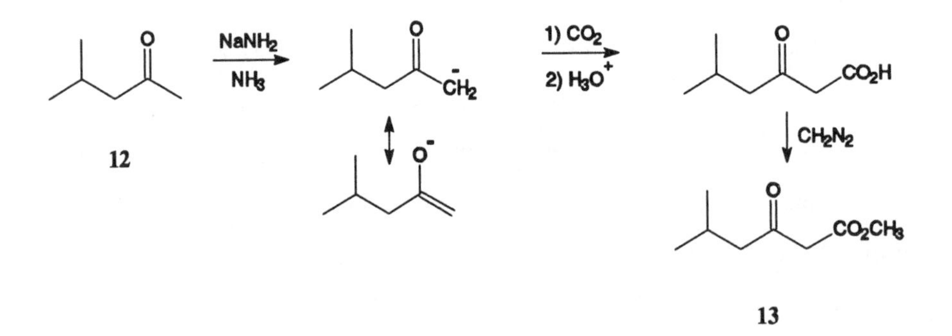

Base-promoted enolization occurs on the methyl group, the less-substituted α carbon atom.

9. Gutsche, C. D.; Redmore, D.; Buriks, R. S.; Nowotny, K.; Grassner, H.; Armbruster, C. W. *J. Am. Chem. Soc.* **1967**, *89*, 1235.

   The reaction is subject to general base catalysis, as evidenced by the linear correlation of log k with pK_{BH^+}. However, there is also a steric component to the catalysis, since there are separate correlations for pyridine bases with 0, 1 and 2 (one compound only) methyl groups ortho to the pyridine nitrogen.

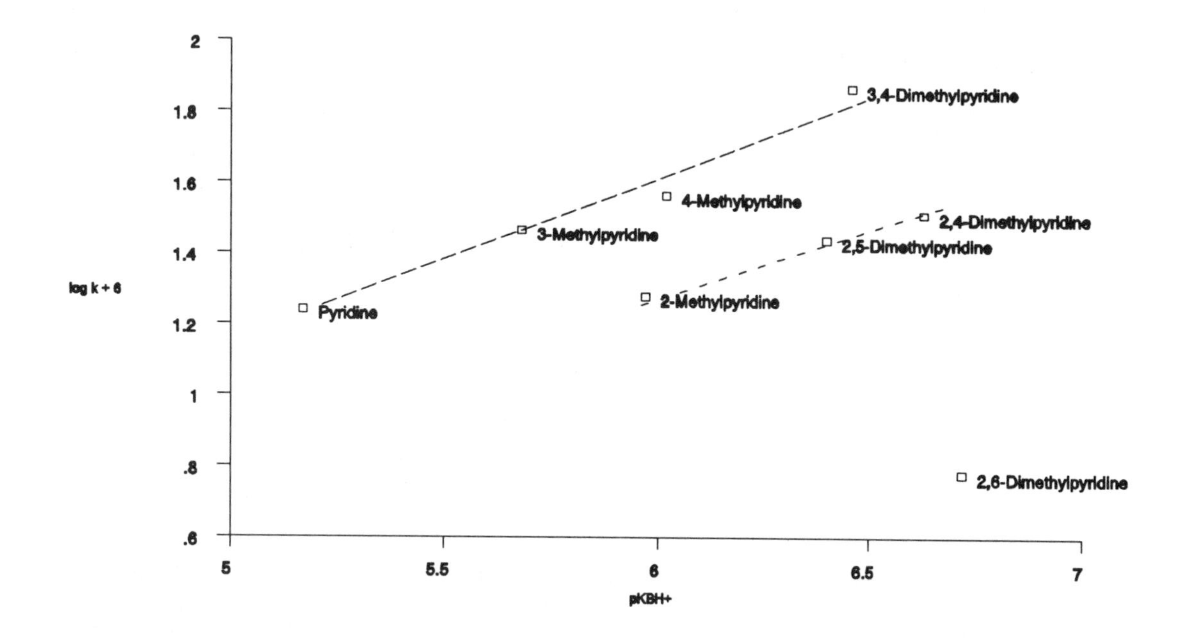

10. Smith, W. T.; McLeod, G. L. *Org. Syn. Coll. Vol. IV* **1963** 345. Also see Walker, J.; Wood, J. K. *J. Chem. Soc.* **1906**, *89*, 598.

The reaction is a haloform reaction.

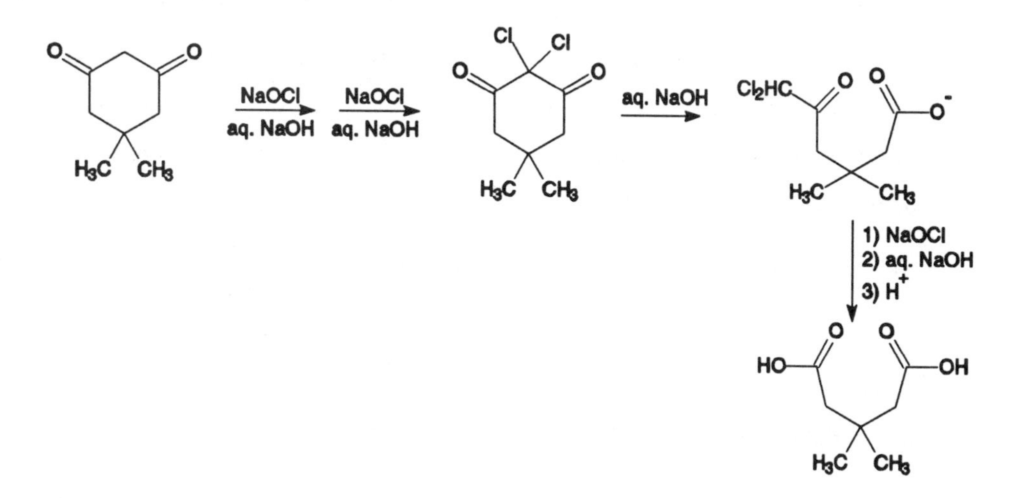

11. Stefanidis, D.; Cho, S.; Dhe-Paganon, S.; Jencks, W. P. *J. Am. Chem. Soc.* **1993**, *115*, 1650. The formate ester should form the more stable tetrahedral intermediate for reasons similar to those for the more stable addition product for hydration of an aldehyde than a ketone.

12. Sørensen, P. E.; Jencks, W. P. *J. Am. Chem. Soc.* **1987,** *109,* 4675.

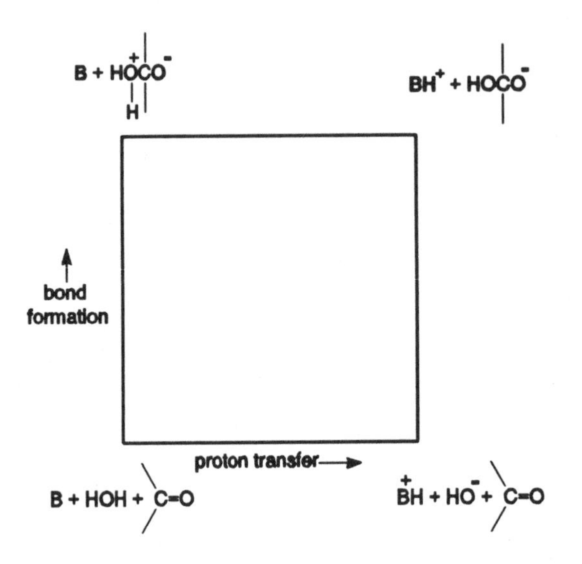

a. A stronger base raises the energy of the left-hand side of the diagram. As a result, the location of the transition structure on the diagram moves toward the bottom of the figure. This movement is the result of movement toward the lower right corner (that is, perpendicular to the reaction coordinate) and toward the lower left corner (a Hammond effect) resulting from a more exothermic reaction).

b. Substitution of better electron donating groups (as in changing from formaldehyde to acetaldehyde, for example), would stabilize the lower left and right corners of the diagram relative to the upper left and right corners. The net effect of the resulting movement parallel to and perpendicular to the reaction coordinate is to move the location of the transition structure to the right on the diagram. For a more detailed discussion of these effects, see page 4687 of Sørensen, P. E.; Jencks, W. P. *J. Am. Chem. Soc.* **1987,** *109,* 4675).

13. Bell, R. P.; Baughan, E. C. *J. Chem. Soc.* **1937,** 1947.

Intramolecular reaction converts two molecules of dihydroxyacetone into a dihemiketal structure.

14. Hurd, C. D.; Saunders, Jr., W. H. *J. Am. Chem. Soc.* **1952,** *74,* 5324.

Intramolecular hemiacetal formation reduces the concentration of carbonyl groups in solution:

For hydroxyaldehydes having the formula $HO(CH_2)_nCHO$, the fraction of free aldehyde groups varies with the chain size because 5- and 6-membered rings are formed more readily than are other size rings. The fraction of free aldehyde varies with *n* as follows: 3, 0.114; 4, 0.061; 5, 0.85; 7, 0.80; 8, 0.91.

15. Drumheller, J. D.; Andrews, L. J. *J. Am. Chem. Soc.* **1955**, *77*, 3290.
    The alcohol should be recovered with the same configuration as in the acetal, since the bond from oxygen to the chiral center is not broken in the hydrolysis reaction.

16. See the discussion in Bender, M. L.; Chen, M. C. *J. Am. Chem. Soc.* **1963**, *85*, 30 (esp. p. 36); *ibid.*, 37 and references therein.
    Steric hindrance makes attack of a nucleophile on the carbonyl carbon atom slower than the dissociation of the acylium ion.

17. Reimann, J. E.; Jencks, W. P. *J. Am. Chem. Soc.* **1966**, *88*, 3973.

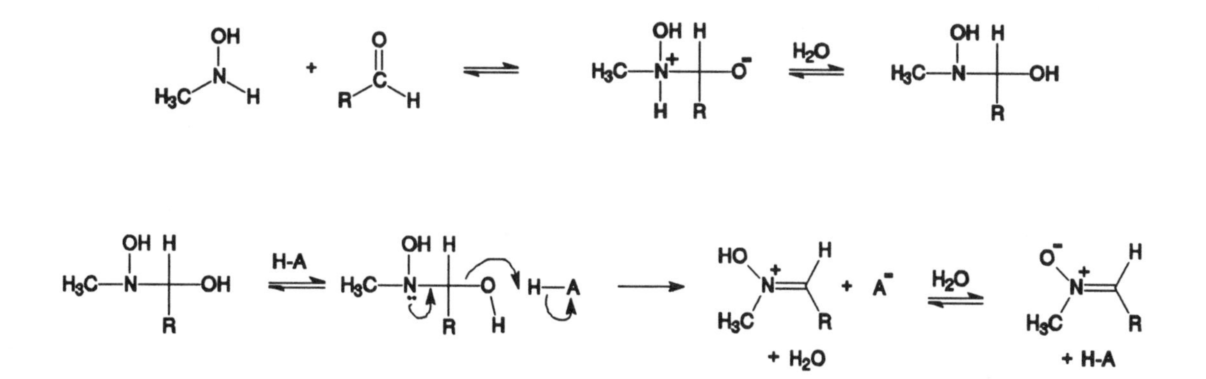

Note the general acid catalysis in the elimination of water in the second step.

18. Olson, A. R.; Youle, P. V. *J. Am. Chem. Soc.* **1951**, *73*, 2468.

Carbonate ion produces general base catalysis for a pathway involving breaking of the acyl carbon — ß-oxygen bond. Acetate ion produces nucleophilic catalysis of a pathway involving breaking the bond between the ß-carbon atom and the oxygen atom.

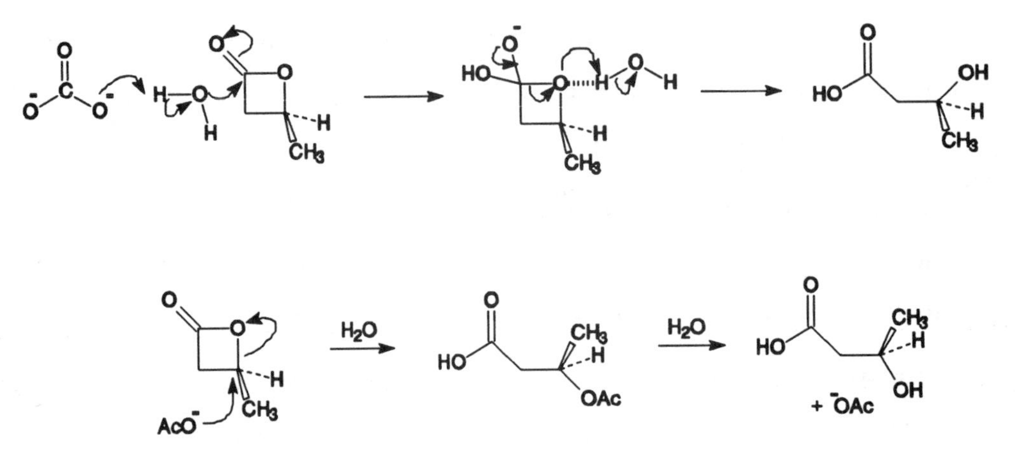

19. Hoz, S.; Livneh, M.; Cohen, D. *J. Org. Chem.* **1986**, *51*, 4537.

Plot log k vs. log K (that is, -pK_a). A good linear correlation is observed, with α = 0.96.

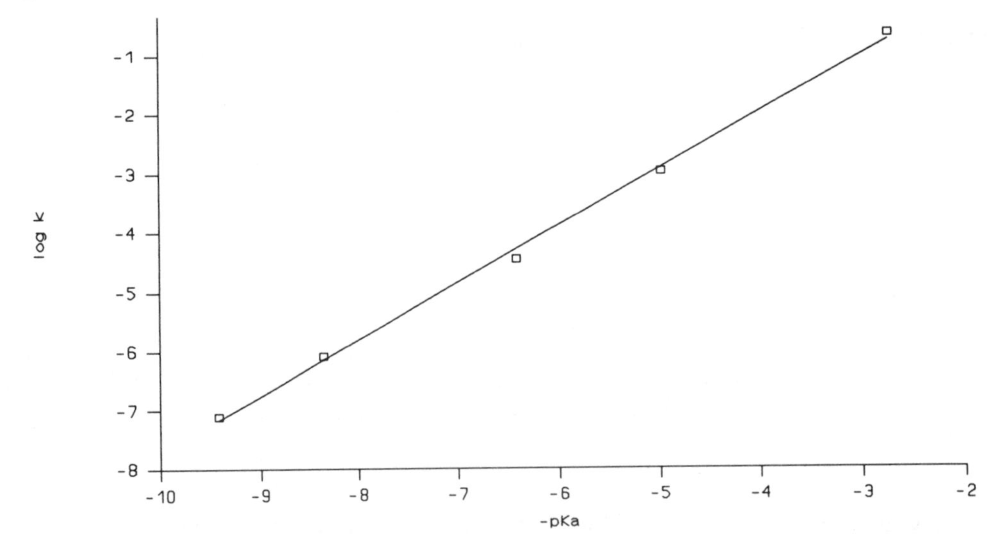

20. Zaugg, H. E.; Papendick, V.; Michaels, R. J. *J. Am. Chem. Soc.* **1964**, *86*, 1399; see also Johnson, S. L. *Adv. Phys. Org. Chem.* **1967**, *5*, 237.

The observation of **19** supports the existence of a tetrahedral intermediate in the reaction. Both products are formed by intramolecular S_N2 reactions. The major product is formed by an S_N2 reaction of the phenol attacking the alkyl bromide moiety after conversion of the lactone to the amide. The minor product is formed by attack of the alkoxide moiety of the tetrahedral intermediate of the acyl substitution reaction with the alkyl bromide group.

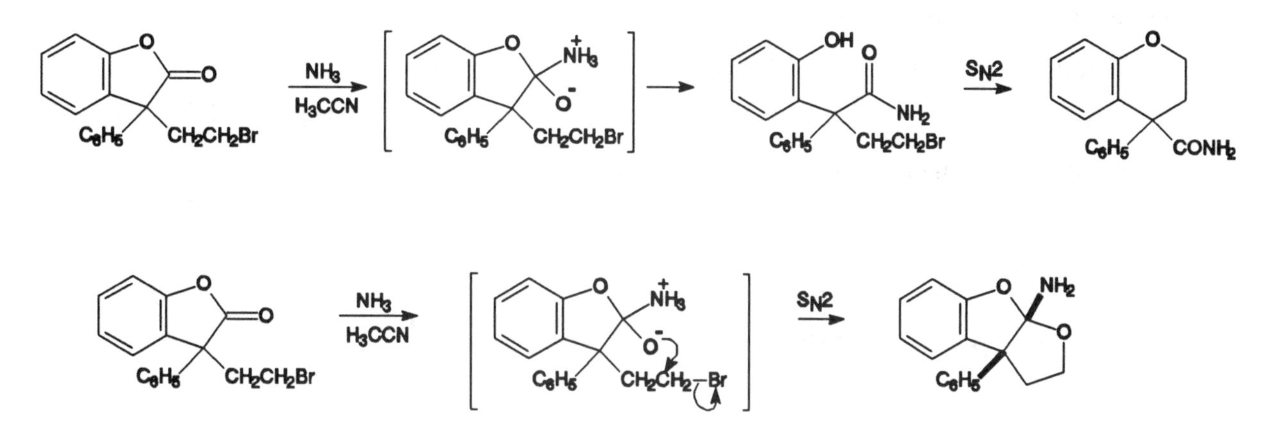

21. Hauser, C. R.; Adams, J. T. *J. Am. Chem. Soc.* **1944**, *66*, 345.

The acylation occurs on the more highly-substituted α carbon atom, consistent with a mechanism involving BF_3-catalyzed enolization. The more stable enol is the one with the more highly-substituted double bond. However, as the size of the R group increases, the rate of reaction of this enol decreases. Therefore, reaction of the minor enol, formed by proton removal from the the methyl group, becomes competitive.

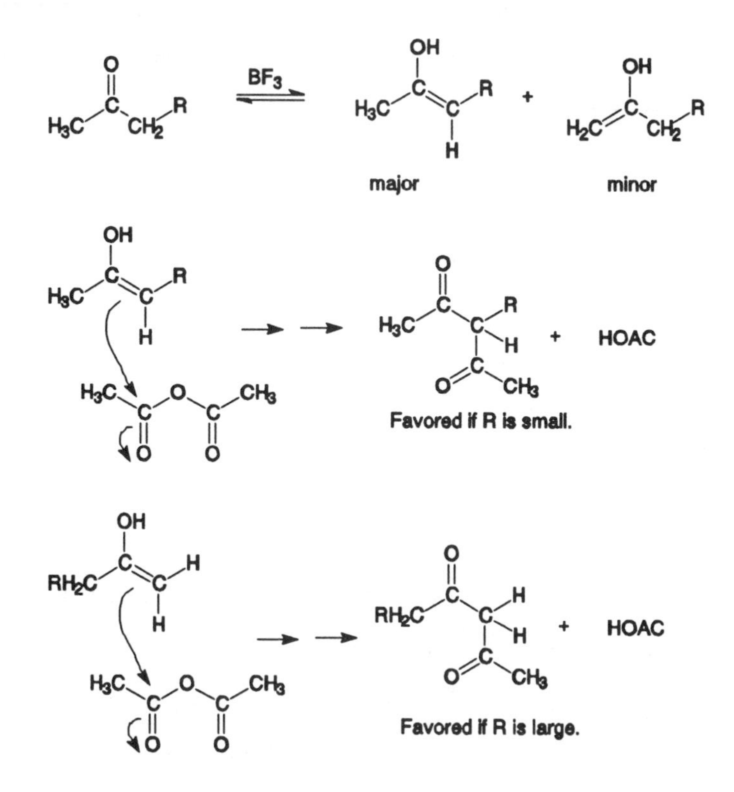

22. (a) Long, F. A.; Pritchard, J. G. *J. Am. Chem. Soc.* **1956**, *78*, 2663; (b) Pritchard, J. G.; Long, F. A. *J. Am. Chem. Soc.* **1956**, *78*, 2667.

The data are consistent with an A-1 mechanism for the reaction in which protonated epoxide is converted to a carbocation in the rate-limiting step for the reaction.

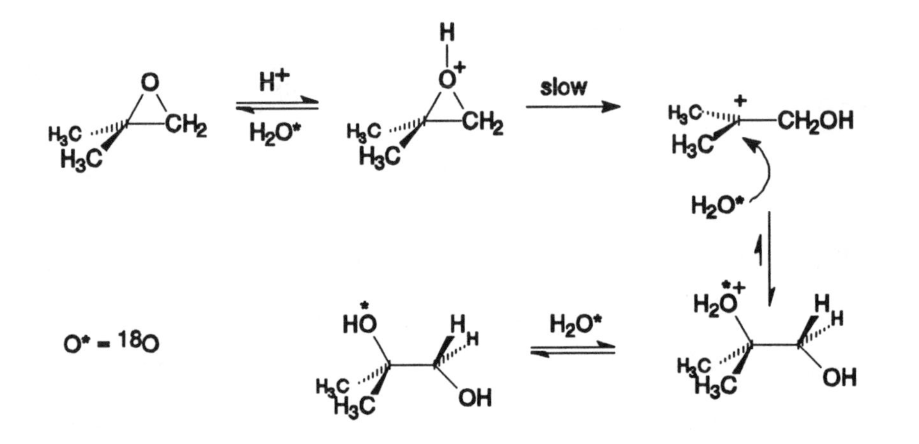

Because 1° carbocations are so much less stable than 2° or 3° carbocations, it seems likely that the minor products in the acid-catalyzed hydrolyses do not arise from opening of the protonated epoxide to 1° carbocations. Instead, attack of water on protonated epoxide may occur in competition with opening of epoxide to carbocation.

Chapter 8

1. Ingold, C. K. *Structure and Mechanism in Organic Chemistry*, 2nd ed.; Cornell University Press: Ithaca, 1969; pp. 457 ff. See especially the table on page 460.

   a) Type III: large decrease. (The products of the reaction are CH_3I and $(CH_3)_3N$).

   b) Type IV: small decrease. (The products of the reaction are $CH_3N^+H_3$ and $(CH_3)_2S$).

2. There is no literature reference for this problem.

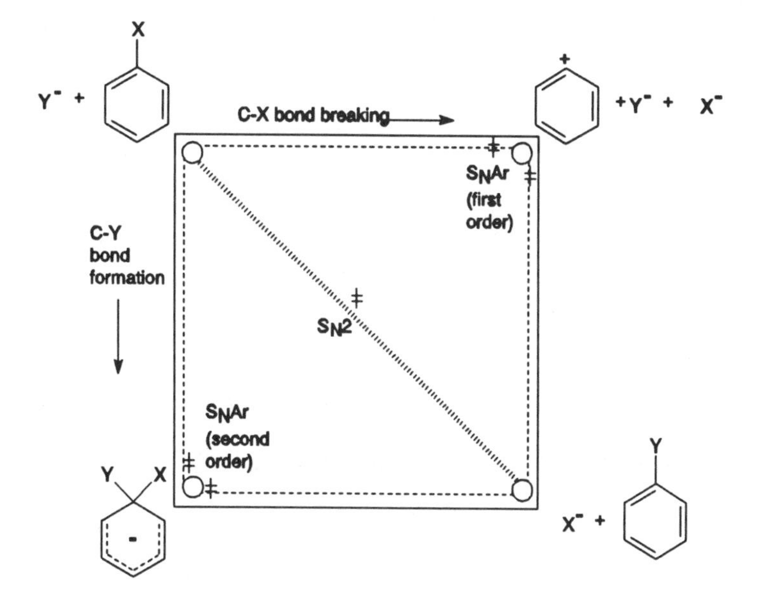

3. Magid, R. M.; Welch, J. G. *J. Am. Chem. Soc.* **1968**, *90*, 5211.

   The result suggest that about 75% of the product is formed by the S_N2' pathway.

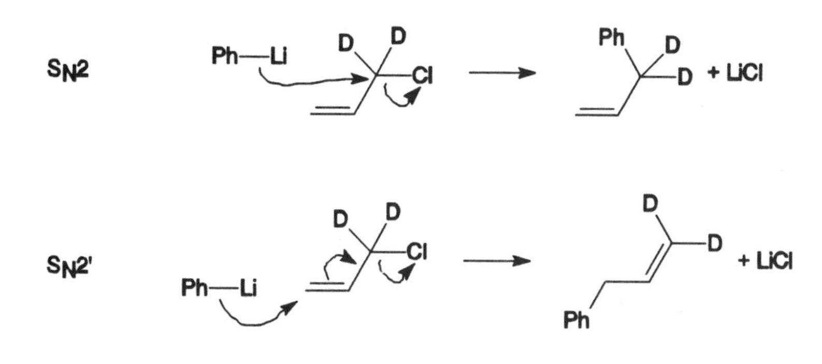

(For a discussion of other possible pathways for the formation of the rearranged product, see the reference cited.)

4. Smith, M. B.; Hrubiec, R. T.; Zezza, C. A. *J. Org. Chem.* **1985**, *50*, 4815.

One product is formed by direct S_N2 reaction, while the other is formed by an S_N2' process. The larger the R group, the greater the steric barrier for the S_N2 pathway, so the greater is the yield of product formed by the S_N2' process. The S_N2' product is obtained in 8% yield when R is methyl, but 88% when R is *t*-butyl.

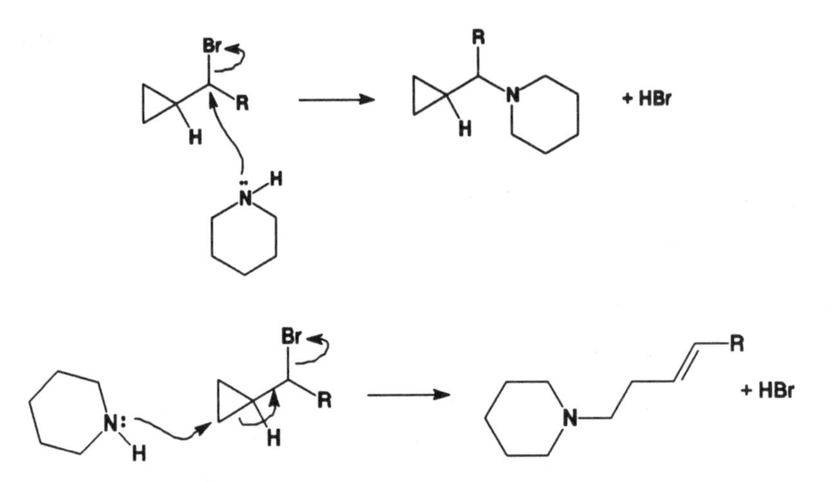

5. The experimental data were reported by Cowdrey, W. A.; Hughes, E. D.; Ingold, C. K. *J. Chem. Soc.* **1937**, 1208. The explanation was proposed by Winstein, S.; Lucas, H. J. *J. Am. Chem. Soc.* **1939**, *61*, 1576. See also Hine, J. *Physical Organic Chemistry*, 2nd ed.; McGraw-Hill: New York, 1962; p. 143. The paper by Cowdrey et. al reported only the optical activity of the reactants and products. The absolute stereochemistry of these species is shown by Klyne, W.; Buckingham, J. *Atlas of Stereochemistry*, 2nd Ed., Vol. 1; Oxford University Press: New York, 1978, p. 5.

The reaction with hydroxide ion is an S_N2 reaction.

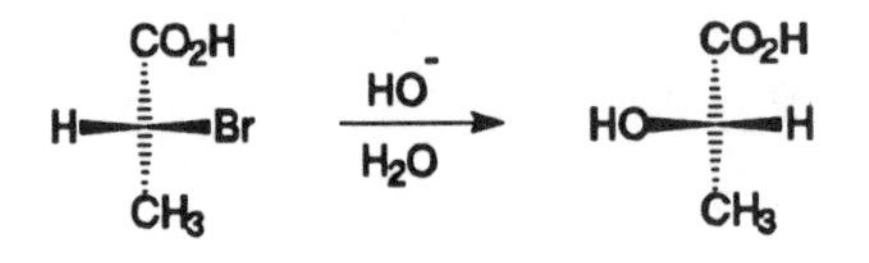

The reaction in water results from anchimeric assistance by the carboxylic acid moiety (shown here as a reaction with the carboxylate ion) in the rate-limiting step of the reaction:

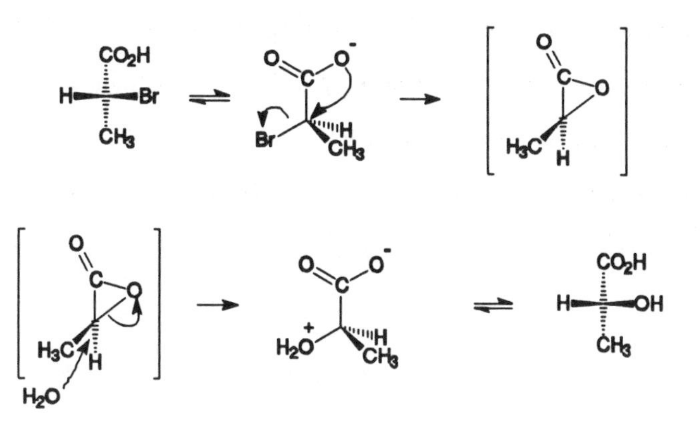

6. McKenzie, A.; Clough, G. W. *J. Chem. Soc.* **1910**, *97*, 2564.
The reaction with $SOCl_2$ is an S_Ni reaction that proceeds with retention of configuration. The reaction with PCl_5 involves S_N2 reaction of Cl^- with the initial product of reaction of the alcohol with PCl_5, so this reaction proceeds with inversion of configuration. For a discussion, see McRae, W. *Basic Organic Reactions*; Heyden & Son, Ltd.: London, 1973; pp. 115-116.

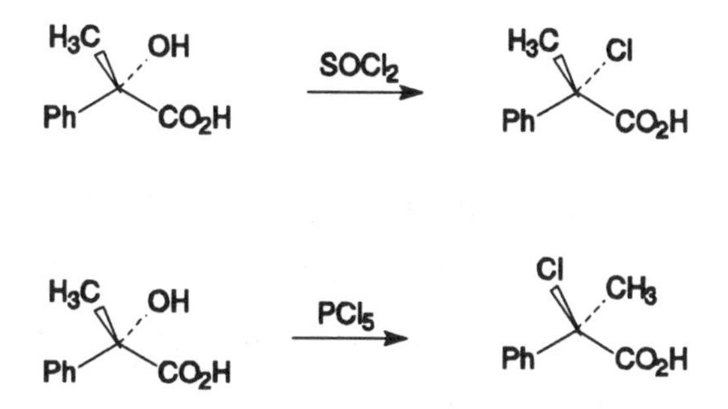

7. Brown, R. F.; van Gulick, N. M. *J. Org. Chem.* **1956**, *21*, 1046.
The rate-limiting step is anchimeric assitance by the NH_2 group in the departure of the bromide ion. The larger the R groups, the greater the percentage of conformations of the reactant in which the amino and CH_2Br groups have the proper orientation for neighboring group participation.

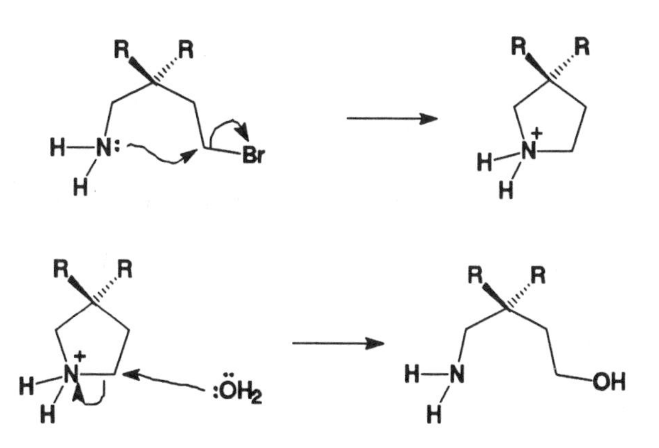

8. a. Cram, D. J. *J. Am. Chem. Soc.* **1949**, *71*, 3863. Also see Gould, E. S. *Mechanism and Structure in Organic Chemistry*; Holt, Rinehart and Winston: New York, 1959; p. 576 and references therein.

The stereochemical designations are evident in the Fischer projections for the reactant and products. For a review of the terminology, see the discussion in Chapter 2.

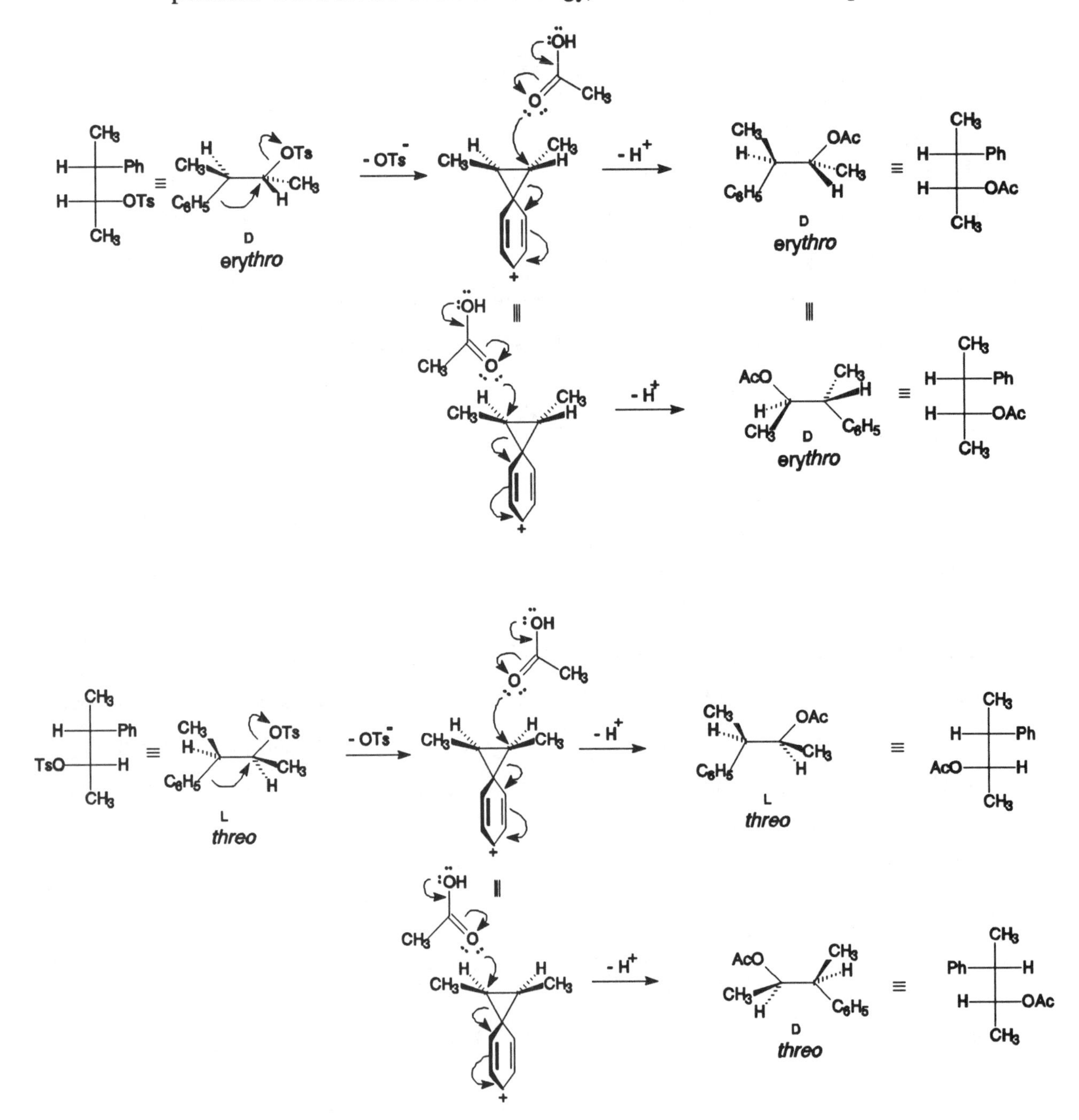

Figure 8.28

b. Cram, D. J. *J. Am. Chem. Soc.* **1949**, *71*, 3875.

The phenonium ion in this case does not have a plane of symmetry because one carbon atom bears a methyl group and the other carbon atom has an ethyl group. However, both carbon atoms are susceptible to nucleophilic attack by solvent.

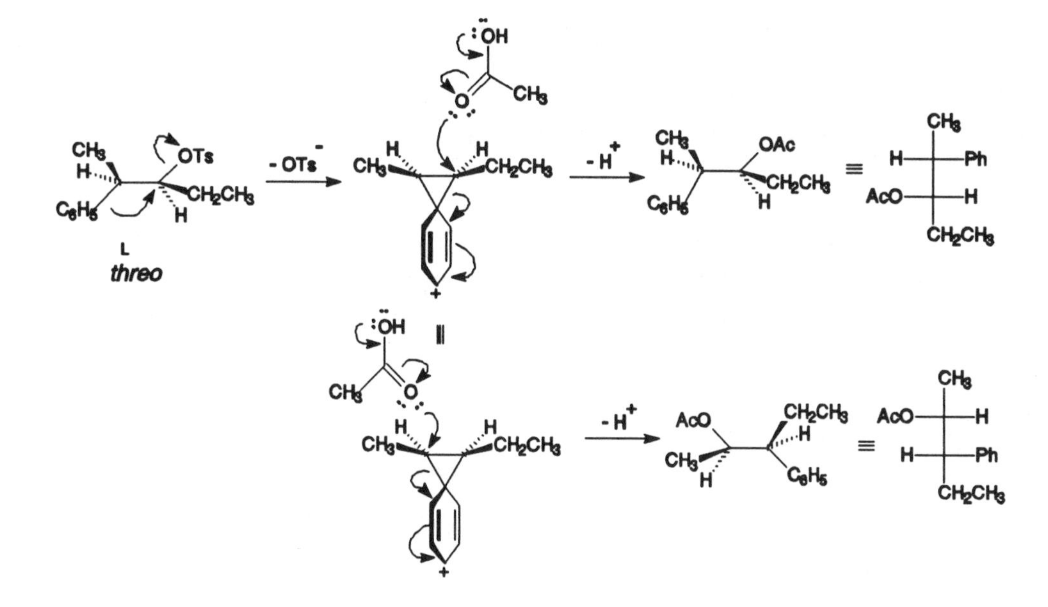

A product mixture that is the mirror image to that formed above would be obtained by reaction of a phenonium ion that is the mirror image of that shown above. Therefore, the reactant that would lead to that phenonium ion is the isomer of 3-phenyl-2-pentyl tosylate in which the methyl and ethyl groups reverse positions from those in L-*threo*-2-phenyl-3-pentyl tosylate.

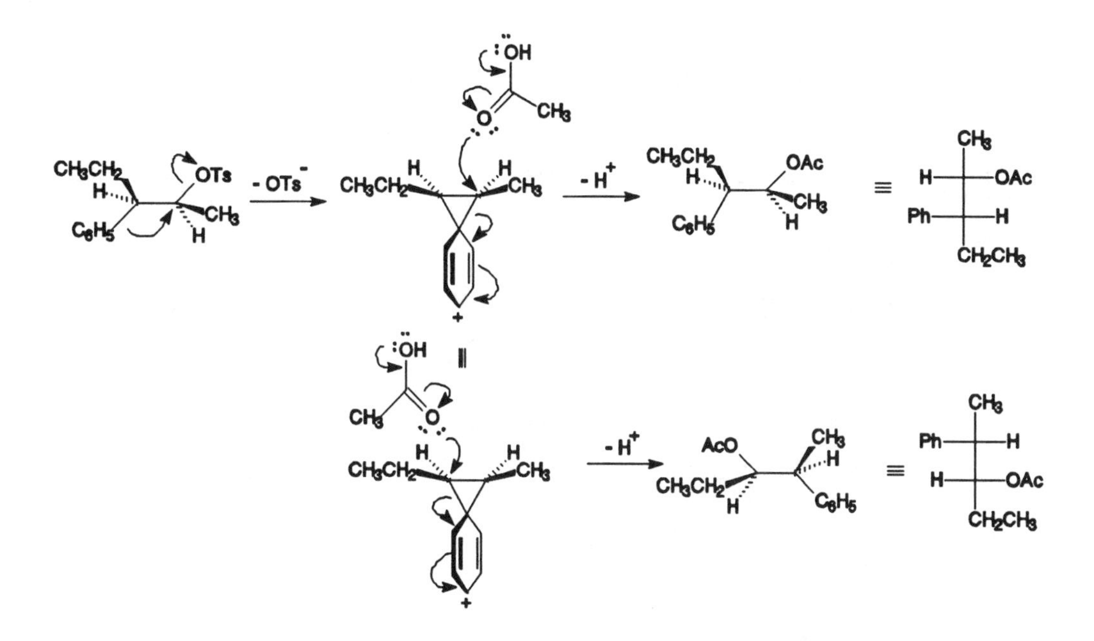

9. There is no literature reference for this problem.

The plot for hexane should resemble that for the gas phase; however the energy should be slightly lower on the right side of the plot. Similarly, the plot for acetone solution should resemble that for aqueous solution but should be slightly higher in energy on the right side of the plot.

10. Winstein, S.; Lucas, H. J. *J. Am. Chem. Soc.* **1939**, *61*, 1581.

The observation of (±)-*erythro*-3-bromo-2-butanol suggests that the ester function is cleaved by acid-catalyzed hydrolysis. This leads to the following mechanism:

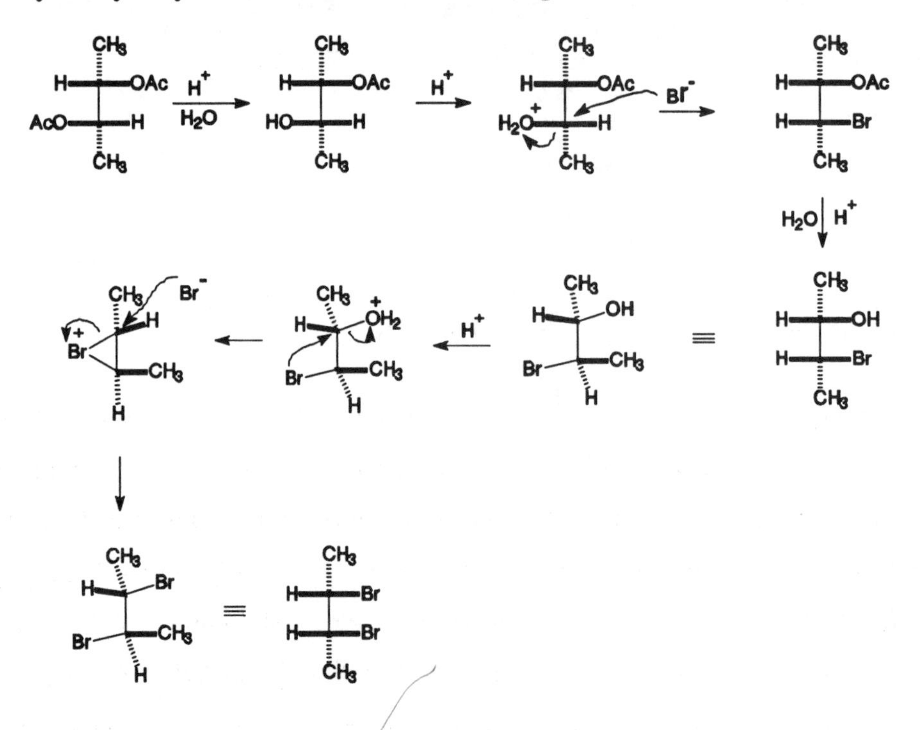

11. The data are from Chapman, J. W.; Strachan, A. N. *Chem. Commun.* **1974**, 293; see also Stock, L. M. *Prog. Phys. Org. Chem.* **1976**, *12*, 21 (data from page 31).

$k_{toluene}/k_{benzene} = (2 \times 41 + 2 \times 2.1 + 51)/6 = 23$ (rounded to the nearest integer).

12. Masci, B. *J. Org. Chem.* **1985**, *50*, 4081.

Partial Rate Factors: Effect of [21C7] on Nitration of Anisole.

	[21C7]			
	0.0 M	0.0082 M	0.032 M	0.131 M
f_{ortho}	5,201	5,009	4,277	3,974
f_{para}	2,917	19,621	56,246	157,651

The results suggest that O_2N^+ ions complexed with 21C7 act as electrophiles in the reaction and that complexed electrophiles are much more sensitive to substituent effects than are uncomplexed nitronium ions. In addition, the relative magnitude of f_p/f_o increases dramatically with [21C7] because of the much greater steric effect of the complexed nitronium ion.

13. Brown, H. C.; Jensen, F. R. *J. Am. Chem. Soc.* **1958**, *80*, 2296.

a. Relative Rate Data for Benzoylation of Benzene and Derivatives in Benzoyl Chloride Solution at 25°.

Compound	k (relative)
Benzene	1
Toluene	110
t-Butylbenzene	72.4

b. $f_o = 30.7$; $f_m = 4.8$ $f_p = 589$

For example, $f_o = 0.03 \times 110 \times 9.3 = 30.7$

c. Plotting log k/k_0 vs. σ^+ gives a good correlation with $\rho = -10.1$. The literature value is -9.57, but the values of σ^+ used in the literature study seem to vary from those listed in chapter 6. (This is particularly true for the value for *p*-Cl, which is not included in this problem.) Note that the term plotted on the y axis is really log p_f, that is log k/k_0 for rates of reaction at *one* position on a benzene ring. The value of k_{rel} for *t*-butyl benzene of 72.4 is multiplied by 6 to determine a f_p of 430 for the *p*-*t*-butyl group (on the assumption that only insignificant yields of ortho and meta products are obtained with such a bulky, ortho,para-directing substituent.

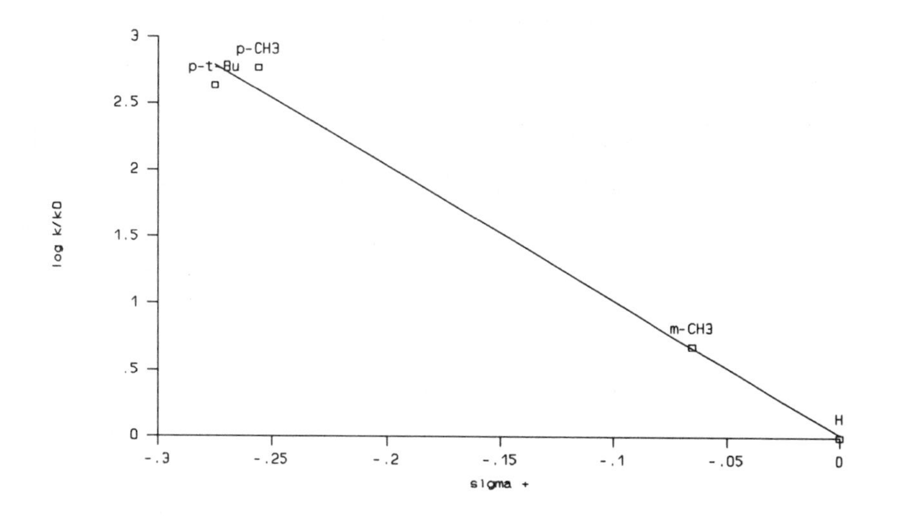

14. a. Himeshima, Y.; Kobayashi, H.; Sonoda, T. *J. Am. Chem. Soc.* **1985,** *107,* 5286.

The reaction is a first-order S_NAr reaction

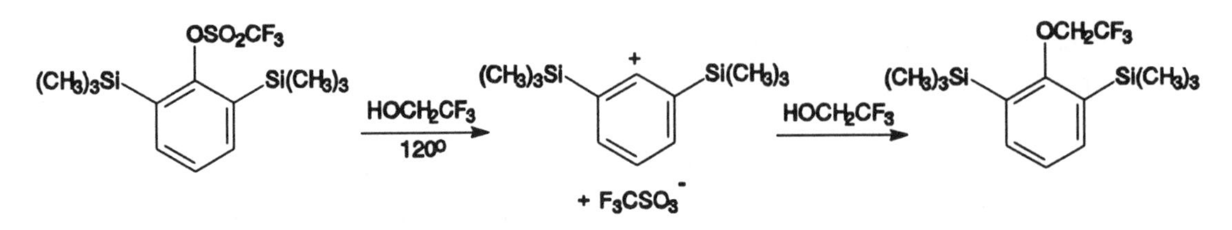

b. The reaction proceeds through formation of a phenyl carbocation which can capture solvent to give the first product or can undergo successive rearrangements involving an ortho *t*-butyl group to give another carbocation, which captures solvent to give the second product.

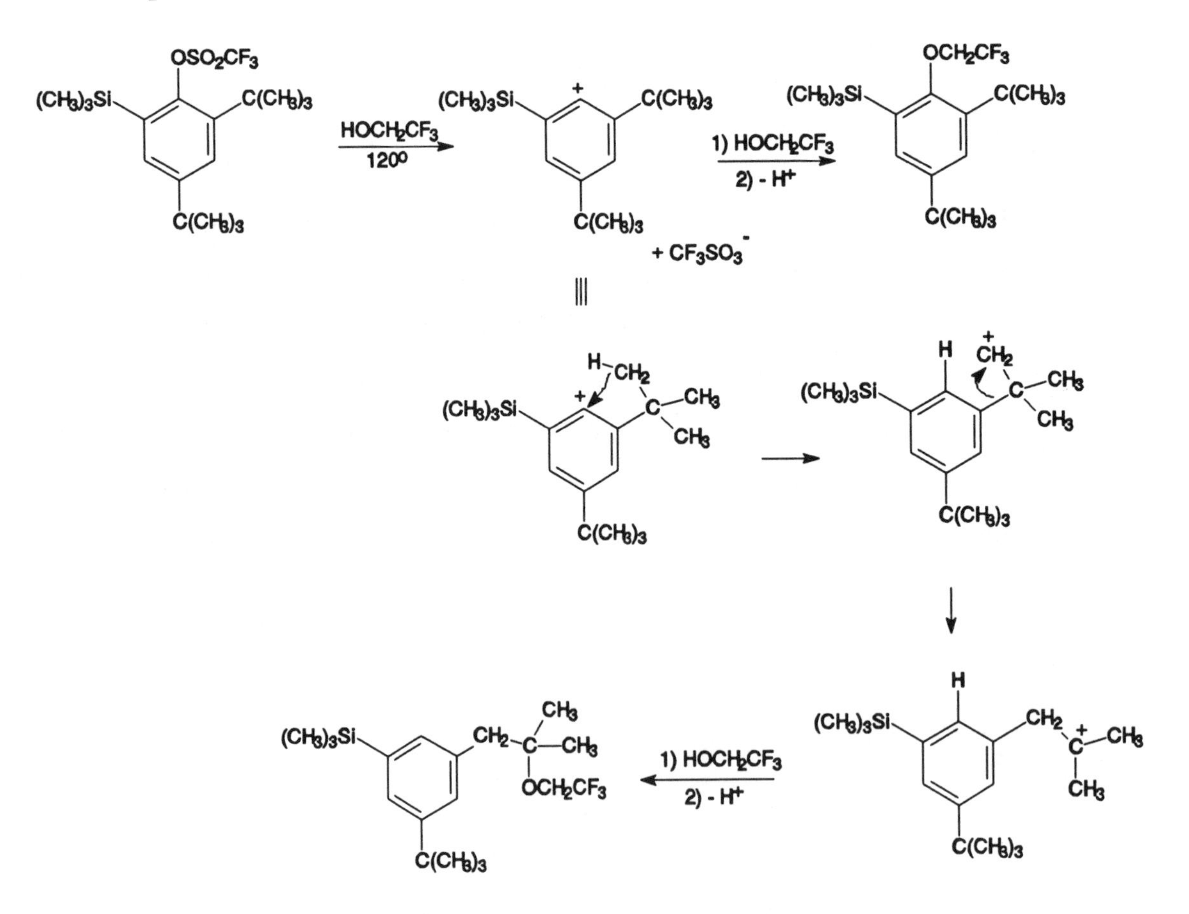

15. Grovenstein, Jr., E.; Kilby, D. C. *J. Am. Chem. Soc.* **1957**, *79*, 2972.

a) 3.97

b) The barrier for formation of the intermediate shown here is lower than that for conversion of the intermediate to product. Therefore, the C-H or C-D bond is broken in the rate-limiting step, and a 1° hydrogen isotope effect is observed. The intermediate is in equilibrium with reactants, so the rate constant for the reaction depends on the concentrations of both phenol and iodine.

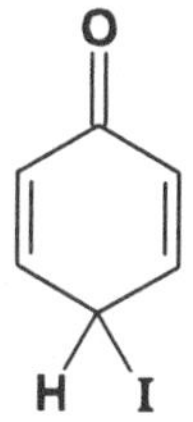

O
H I

c) As noted in the reference, phenols and amines can give intermediates without formal charge; sulfonation gives zwitterionic intermediate with no net formal charge.

16. For a discussion of mechanistic possibilities, see Truce, W. E.; Kreider, E. M.; Brand, W. W. *Org. React.* **1970**, *18*, 99. The accepted mechanism is an S_NAr reaction.

17. The incorporation of protons from solvent was demonstrated by Bunnett, J. F.; Rauhut, M. M.; Knutson, D.; Bussell, G. E. *J. Am. Chem. Soc.* **1954**, *76*, 5755, who found that the product contained deuterium when the reaction was run in D_2O - CH_3CH_2OD. Samuel, D. *J. Chem. Soc.* **1960**, 1318, studied the reaction in $H_2{}^{18}O$ - CH_3CH_2OH and found that the carboxylic acid group incorporated one ^{18}O atom. Rosenblum, M. *J. Am. Chem. Soc.* **1960**, *82*, 3796, studied the reaction with reagents containing ^{15}N. For details of the information obtained from these mechanistic studies, see these and other references cited in the chapter.

18. Roberts, J. D.; Semenow, D. A.; Simmons, Jr., H. E.; Carlsmith, L. A. *J. Am. Chem. Soc.* **1956**, *78*, 601.

This reaction is discussed on page 606 of Roberts, J. D.; Semenow, D. A.; Simmons, Jr., H. E.; Carlsmith, L. A. *J. Am. Chem. Soc.* **1956**, *78*, 601. The experimental observations were reported by Wright, R. E.; Bergstrom, F. W. *J. Org. Chem.* **1936**, *1*, 179. The observation of catalysis by KNH_2 strongly suggests that the mechanism is not an S_NAr substitution but proceeds, instead, by a benzyne mechanism. The KNH_2 reacts with chlorobenzene to form benzyne, which adds the triphenylmethide ion to form tetraphenylmethane.

19. If the mechanism shown operates in this particular reaction, it should be possible to detect *m*-chlorotrifluoromethylbenzene in the reaction mixture. If such a mechanism were to operate in general, then isotopic labeling should show formation of, for example, aniline-*3*-^{14}C from reaction of chlorobenzene-*1*-^{14}C.

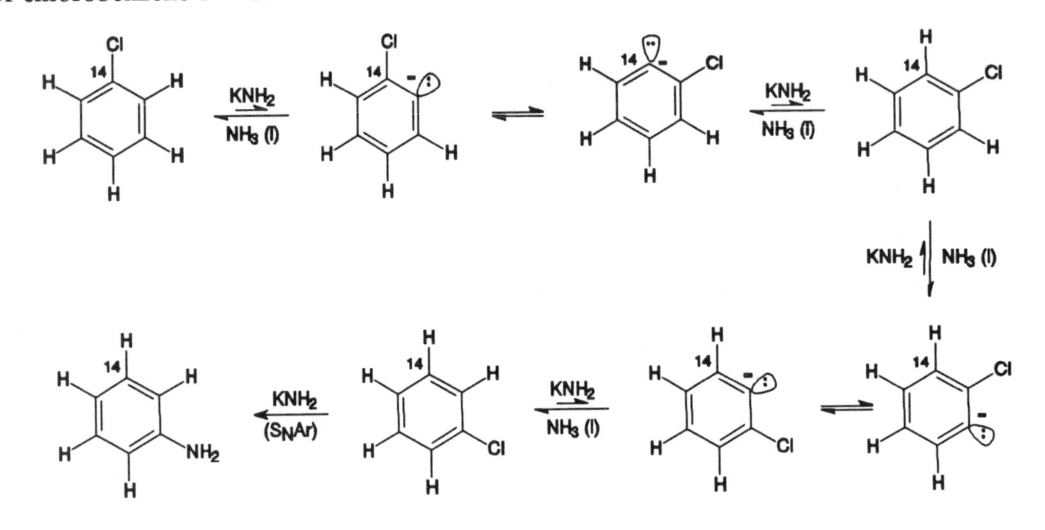

20. Katritzky, A. R.; Laurenzo, K. S. *J. Org. Chem.* **1986**, *51*, 5039.

This is an S_NAr reaction in which the Meisenheimer complex undergoes base-promoted ß-elimination to form a tautomer of the product.

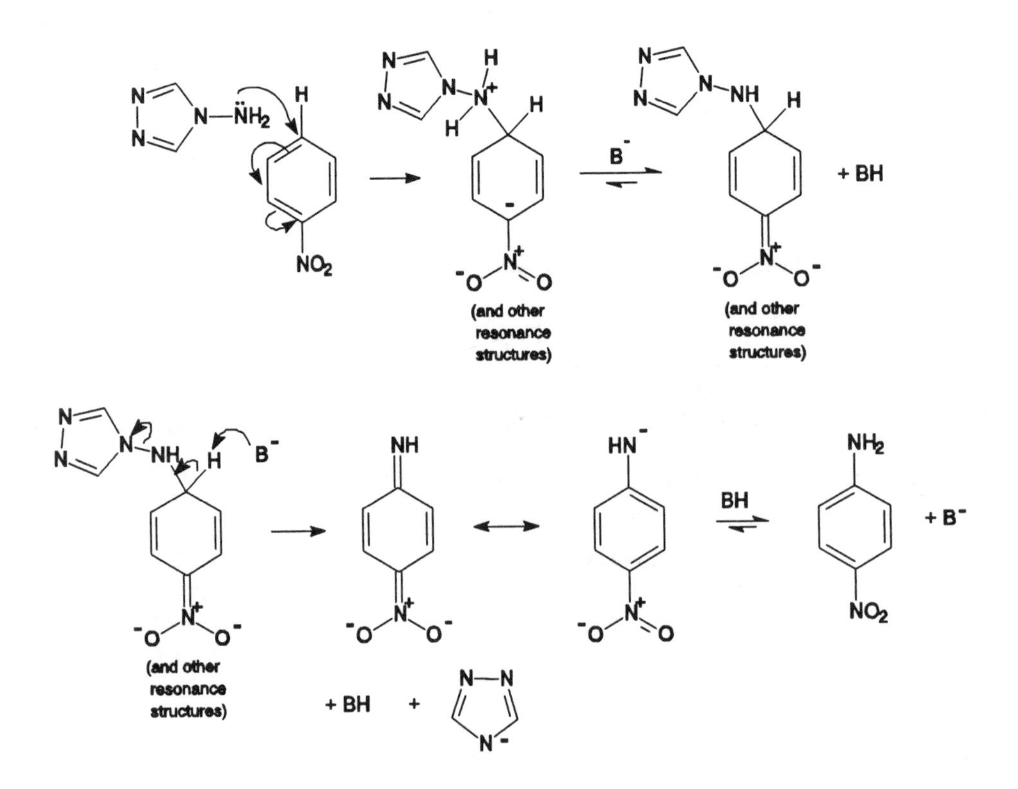

21. Bunnett, J. F.; Garbisch, Jr., E. W.; Pruitt, K. M. *J. Am. Chem. Soc.* **1957**, *79*, 385.

The lack of an element effect indicates that the reaction occurs in two steps: rate-limiting formation of a Meisenheimer complex, followed by elimination of the bromide ion.

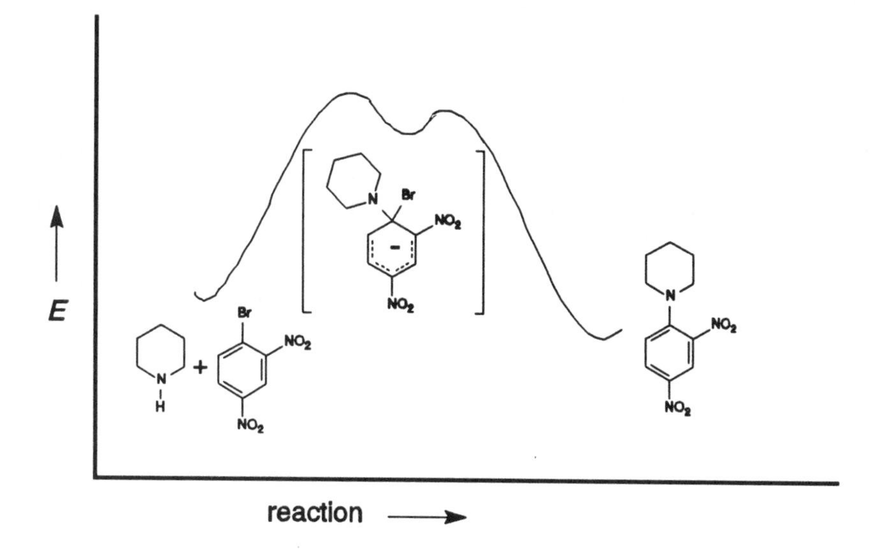

22. For a general discussion, see Roberts, J. D.; Vaughan, C. W.; Carlsmith, , L. A.; Semenow, D. A. *J. Am. Chem. Soc.* **1956**, *78*, 611.

Only one benzyne can be produced. Addition of amide ion to the resulting benzyne gives two phenyl carbanions. The slight preference for *meta* product may result from a slight steric interaction between the amino group and an *ortho* methyl group.

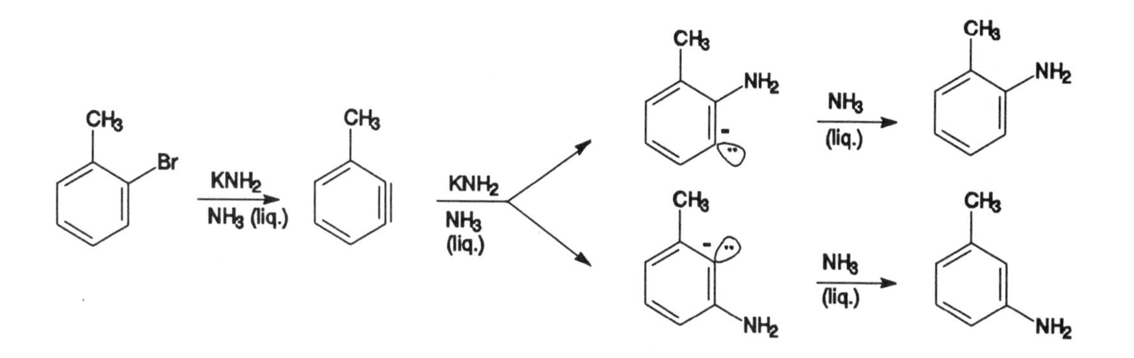

23. Roberts, J. D.; Semenow, D. A.; Simmons, Jr., H. E.; Carlsmith, L. A. *J. Am. Chem. Soc.* **1956**, *78*, 601 (especially p. 603).

a. These compounds are unreactive by the benzyne mechanism because in each case there is not a hydrogen atom ortho to the halogen atom.

b. The lack of reaction of these compounds alone does not establish the benzyne mechanism because in each case the halogen atom is sterically hindered by the ortho substitutents. This steric hindrance could decrease the rate of any reaction step involving addition of a nucleophile to the carbon atom bearing the halogen atom or to an adjacent atom.

Chapter 9

1. Buckles, R. E.; Bader, J. M.; Thurmaier, R. J. *J. Org. Chem.* **1962**, *27*, 4523.
   The meso product is the more stable product and would be expected to be the major product of a mechanism that produced the more stable product no matter what the stereochemistry of the reactant. Thus investigating the reaction of both isomers is necessary to determine whether the reaction is stereospecific or merely stereoselective.

2. Buckles, R. E.; Forrester, J. L.; Burham, R. L.; McGee, T. W. *J. Org. Chem.* **1960**, *25*, 24.
   The product is 2-bromo-3-chloro-3-phenylpropionic acid; the stereochemistry is presumed to be that from anti addition. The regiochemistry is consistent with a mechanistic model in which BrCl adds as Br^+ and Cl^-.

3. Bellucci, G.; Bianchini, R.; Chiappe, C. *J. Org. Chem.* **1991**, *56*, 3067.
   CH_3CN adds as a nucleophile to the bromonium ion; subsequent addition of water, deprotonation, and tautomerization produces the product.

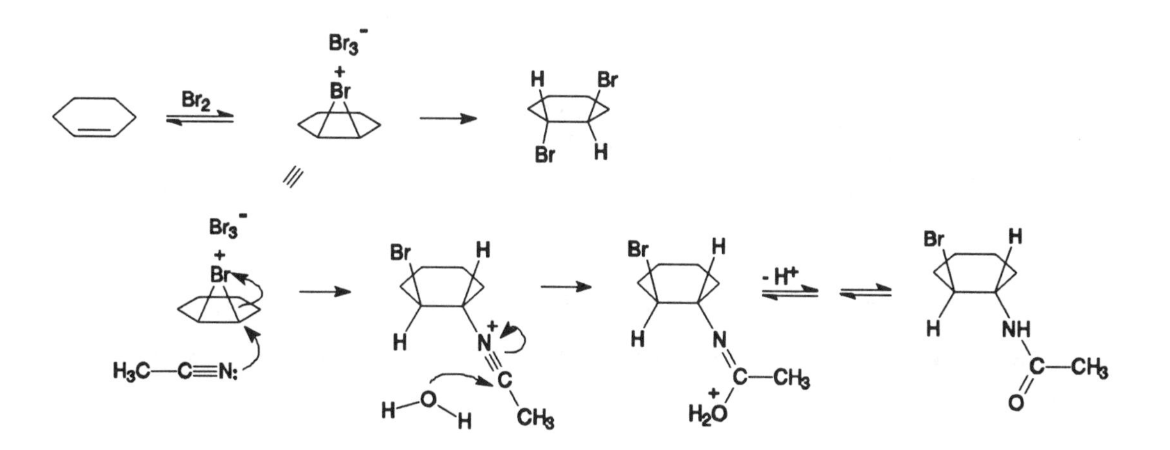

4. Bedoukian, P. Z. *J. Am. Chem. Soc.* **1944**, *66*, 1325

Acetylation of the enol of the reactant leads to the enol acetate (**A**), which adds bromine to form **B**. Methanolysis leads to the formation of dimethyl acetal of α-bromophenylacetaldehyde (**C**).

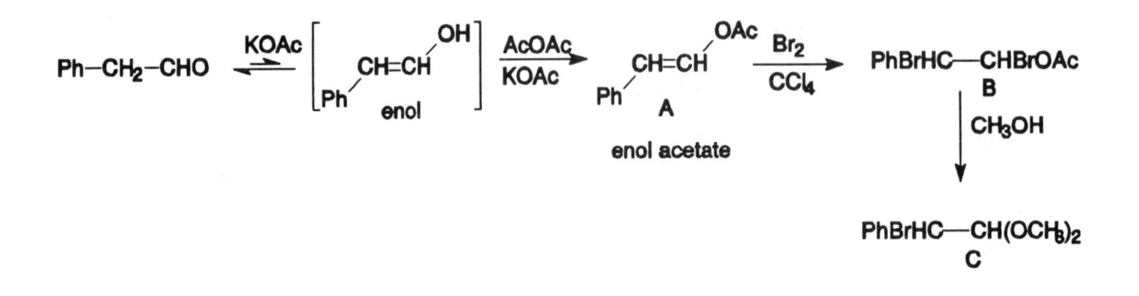

5. Tarbell, D. S.; Bartlett, P. D. *J. Am. Chem. Soc.* **1937**, *59*, 407.

The lactone is formed from the chloronium ion or chlorocarbocation. The observation of diastereomeric products suggests that closure of the chlorocarbocation must be faster than rotation about the carbon-carbon single bond. However, it is unlikely that the chlorolactone is formed in one step through intramolecular reaction of a chloronium ion intermediate because nucleophilic attack of the carboxylate group on the back side of the C-Cl bond is not sterically feasible in the chloronium ion. A mechanism for reaction of the (*E*)-diastereomer is shown here, and the mechanism for the (*Z*)-diastereomer is analogous.

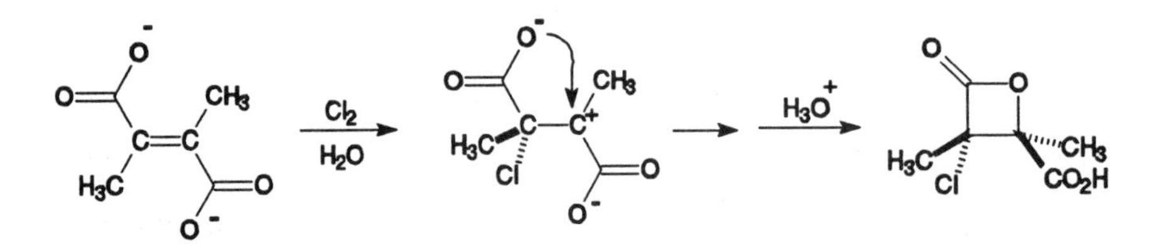

6. Rolston, J. H.; Yates, K. *J. Am. Chem. Soc.* **1969**, *91*, 1477.

Dioxane acts as a nucleophile to produce an adduct (shown below), which can then undergo attack by Br- after rotation about the carbon-carbon single bond.

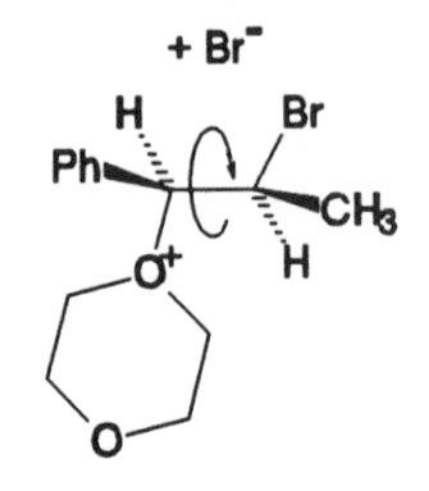

7. van Tamelen, E. E.; Shamma, M. *J. Am. Chem. Soc.* **1954**, *76*, 2315.

For a) - c), an intermediate iodinium ion reacts with a neighboring carboxylate group to form a 5- or 6-membered lactone ring. In d), the iodinium ion does not react with the carboxylate group because formation of a 5- or 6-membered ring is not sterically feasible. Instead, a decarboxylation occurs to give a neutral compound.

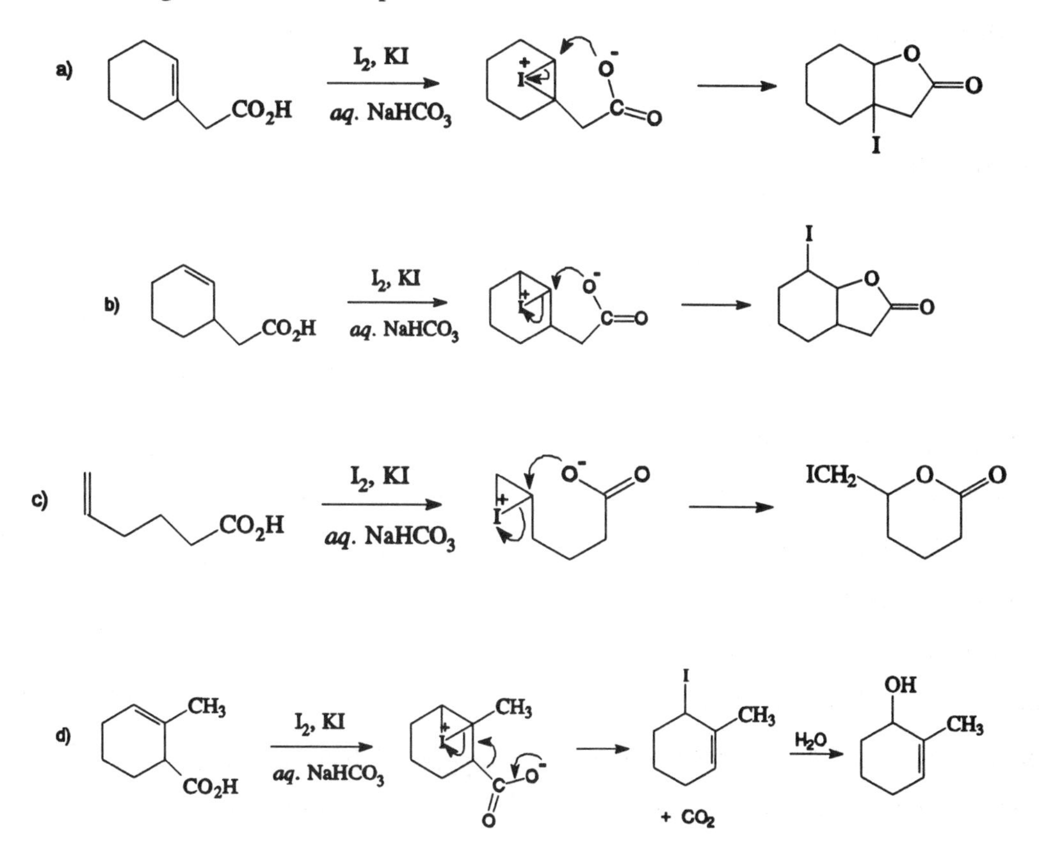

8. Peterson, P. E.; Tao, E. V. P. *J. Am. Chem. Soc.* **1964**, *86*, 4503.

The observation of 5-chloro-2-deuterio-2-hexyl trifluoroacetate, as well as kinetic data not cited in the problem, are consistent with the involvement of a 5-membered chloronium ion:

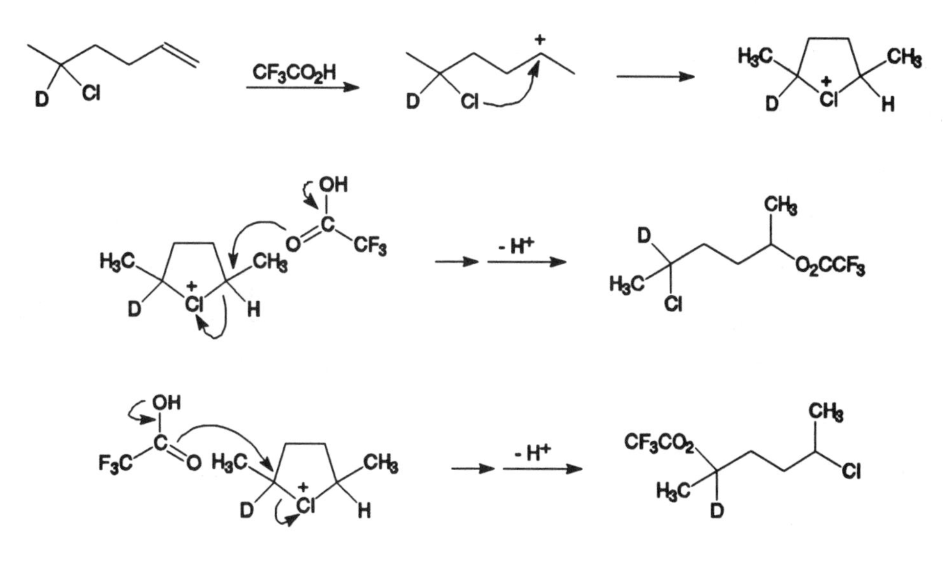

Since the chloronium ion should lead to essentially equal yields of the two products, it appears that about one-third of the 5-chloro-5-deuterio-2-hexyl trifluoroacetate formed by addition of trifluoroacetic acid by ordinary Markovnikov addition (that is, by a pathway not involving the chloronium ion intermediate).

9. Fraenkel, G.; Bartlett, P. D. *J. Am. Chem. Soc.* **1959,** *81*, 5582.
PhĊHCH2I radicals (most likely generated through photolysis of I_2) add to styrene to produce benzylic radicals which dimerize to the product.

$$I_2 \xrightarrow{h\nu} 2\,I\cdot$$

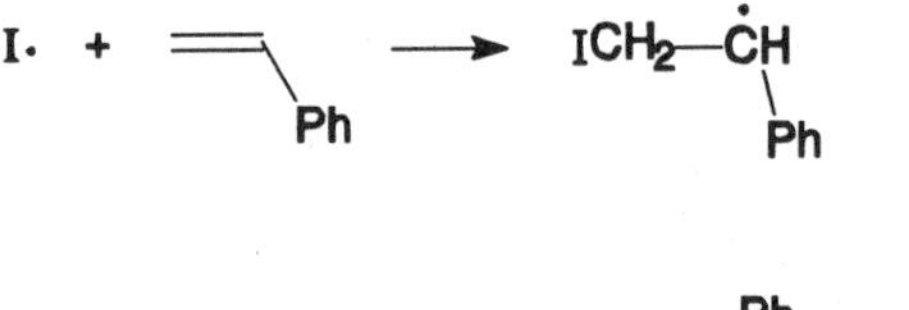

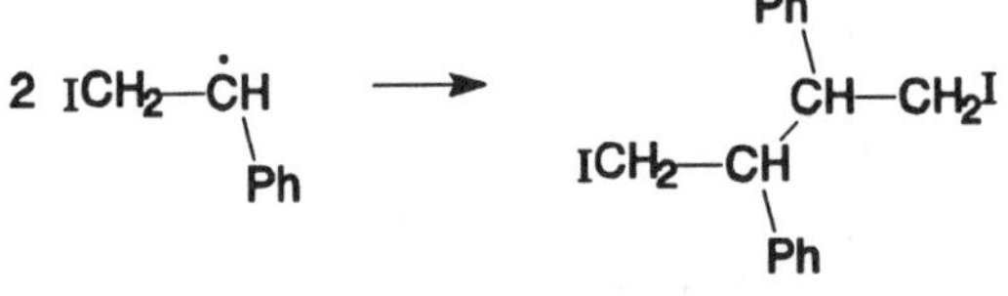

The major product of the reaction is a structure suggesting the combination of I_2 with three styrene molecules. Its formation is conveniently represented as follows:

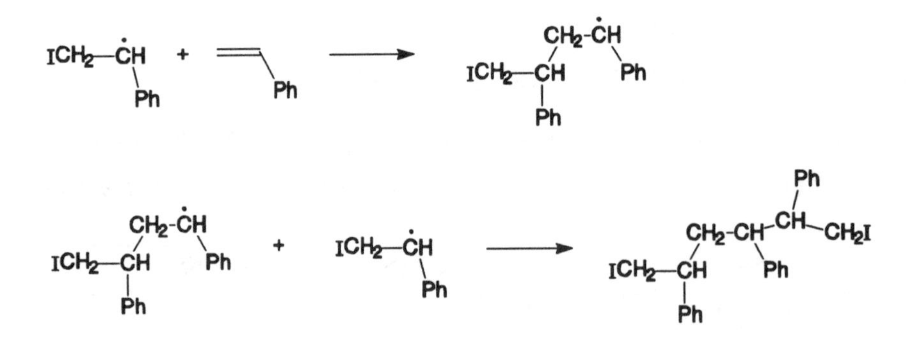

10. Okuyama, T.; Sakagami, T.; Fueno, T. *Tetrahedron* **1973**, *29*, 1503. See also the discussion in Chwang, W. K.; Knittel, P.; Koshy, K. M.; Tidwell, T. T. *J. Am. Chem. Soc.* **1977**, *99*, 3395. Protonation of the terminal methylene group produces a resonance-stabilized carbocation that can undergo attack by water (by one or the other or both of the routes shown) to produce the ketone.

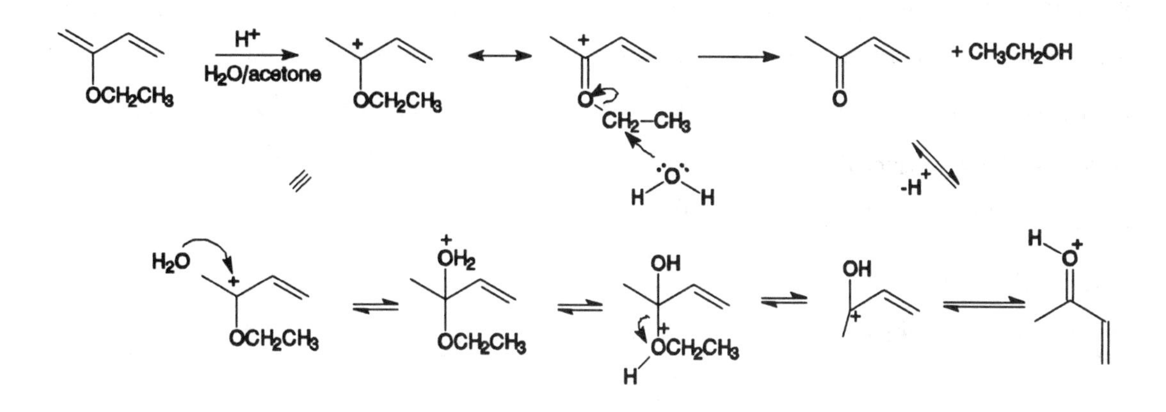

11. Bellucci, G.; Bianchini, R.; Vecchiani, S. *J. Org. Chem.* **1987**, *52*, 3355.
The first two products are formed by attack of Br^- (Br_3^- under the reaction conditions) on a bromonium ion. The last two products arise from attack of Br^- (or Br_3^-) on an intermediate resulting from neighboring group participation of the ester function.

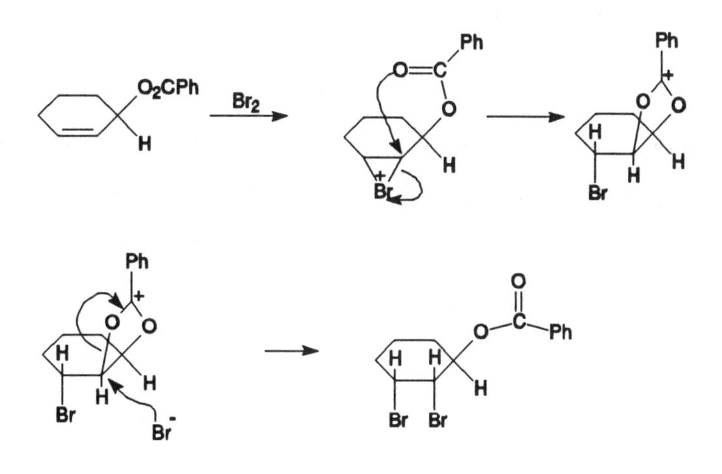

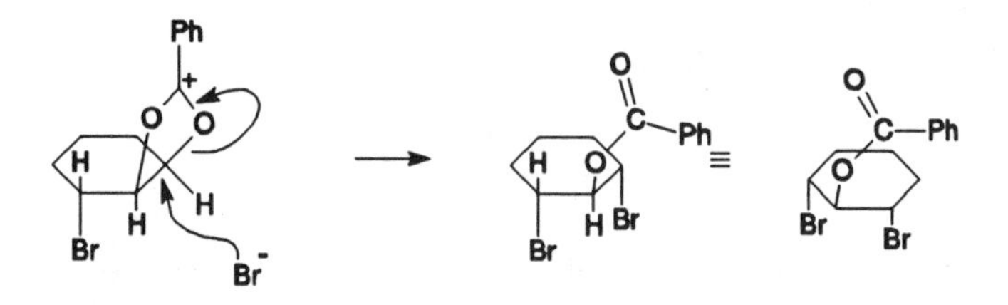

12. Fahey, R. C.; Schneider, H.-J. *J. Am. Chem. Soc.* **1968**, *90*, 4429.

The product distributions correlate with the bridging ability of the adding electrophilic atom. Those atoms that bridge well (such as Br) lead to predominantly anti addition. Those that bridge poorly or not all (such as H) lead to open cations that can rotate before addition of the nucleophile, so a mixture of syn and anti addition is observed.

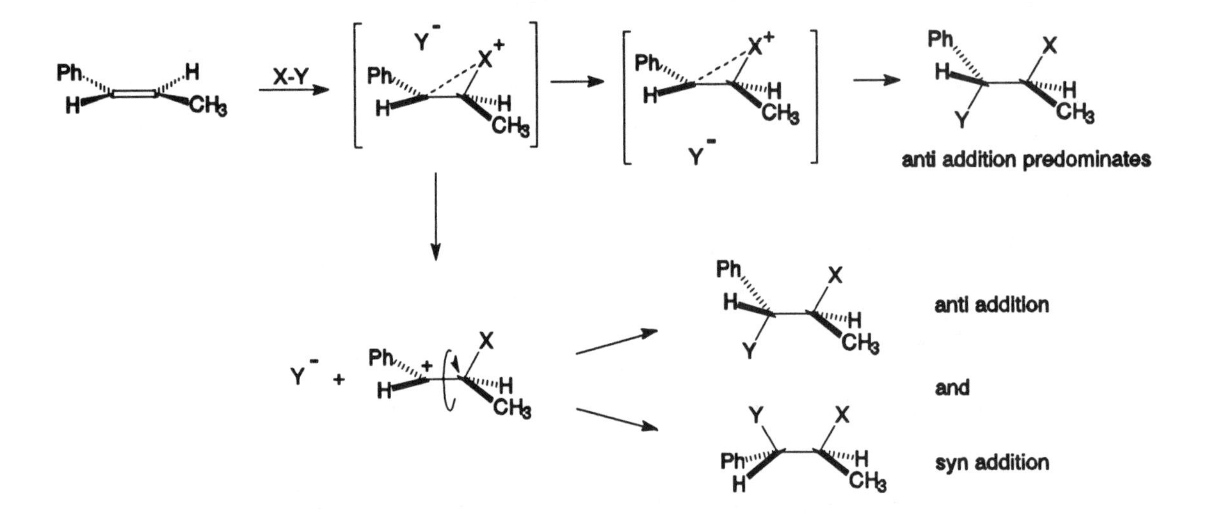

13. Rolston, J. H.; Yates, K. *J. Am. Chem. Soc.* **1969**, *91*, 1469.

Styrene reacts through a benzylic carbocation or bridged cation with considerable benzylic carbocation character. The dimethylstyrene may give a bridged ion or a competition between benzyl and tertiary carbocations. Therefore using *p*-nitro or other electron withdrawing groups in the dimethyl isomer might shift the product distribution even more.

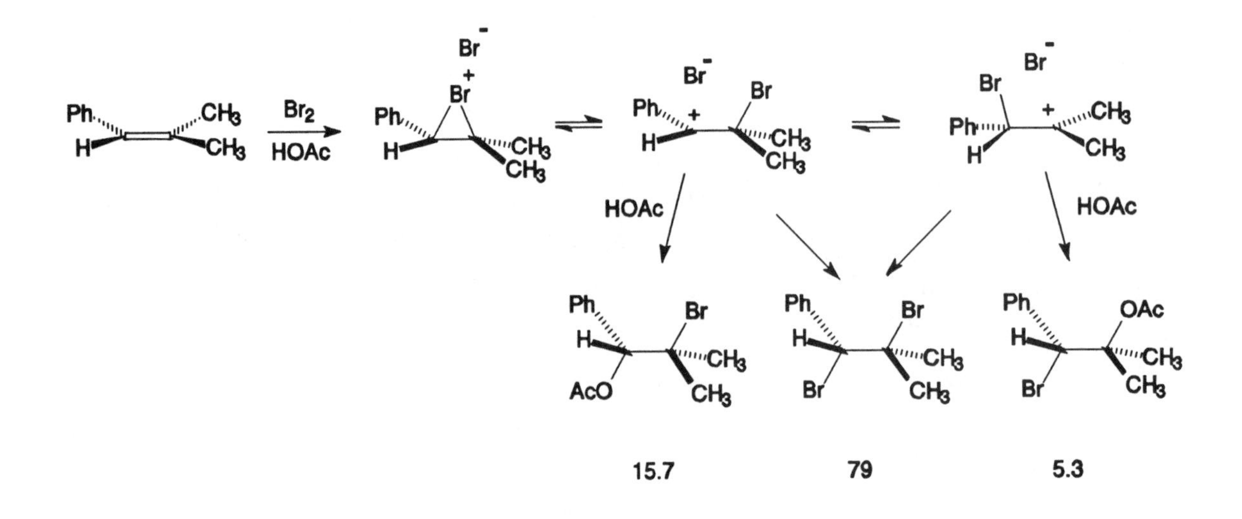

14. Cabaleiro, M. C.; Johnson, M. D. *J. Chem. Soc. B* **1967**, 565.

The reaction proceeds through chlorocarbocation ion pairs **A** and **B** that can undergo conformational change before reaction with Cl^- to form III and IV by syn addition or reaction with solvent to form I and II by anti addition. In the presence of added chloride ion, anti addition of Cl^- is also observed.

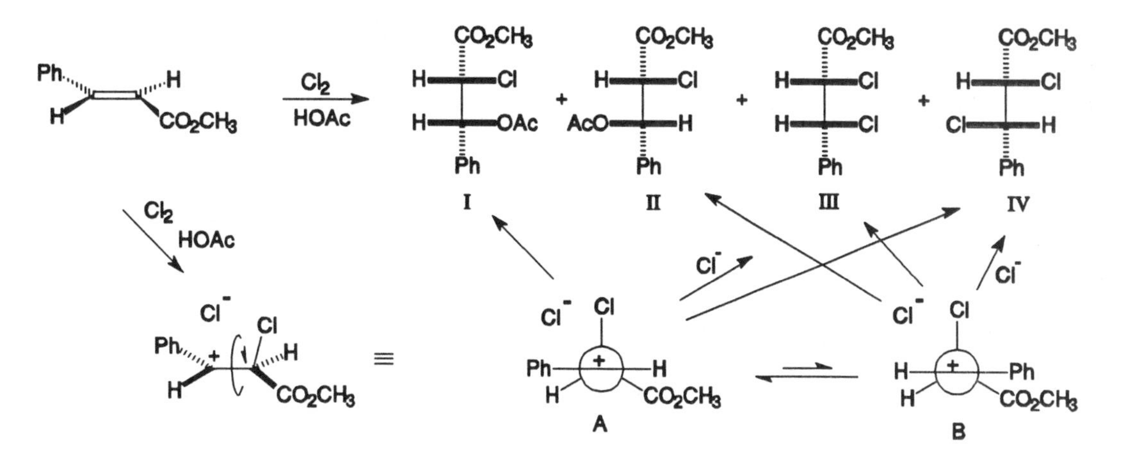

15. Rozen, S.; Brand, M. *J. Org. Chem.* **1986**, *51*, 3607. [See top of left column on p. 3608.]

The product is *threo*-3,4-difluorohexan-1-ol acetate, formed by syn addition of F_2 to the alkene.

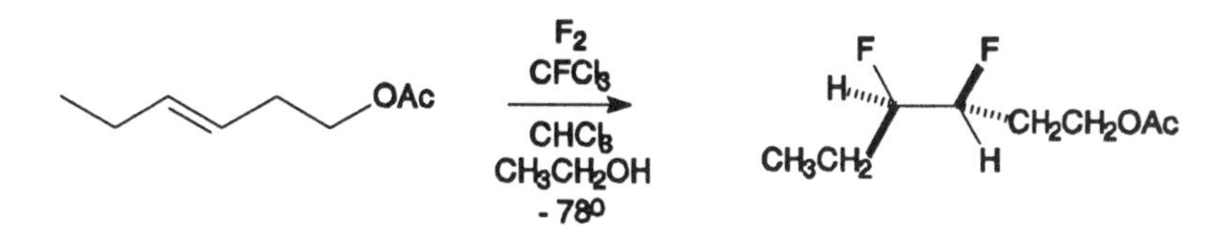

16. Jacobs, T. L.; Searles, Jr., S. *J. Am. Chem. Soc.* **1944**, *66*, 686.

The rate-limiting step in the reaction is protonation of the alkyne. Reaction of the butyoxyalkyne is faster than the corresponding reaction of most other alkynes because in this case the intermediate vinyl carbocation can be stabilized by resonance with the adjacent oxygen.

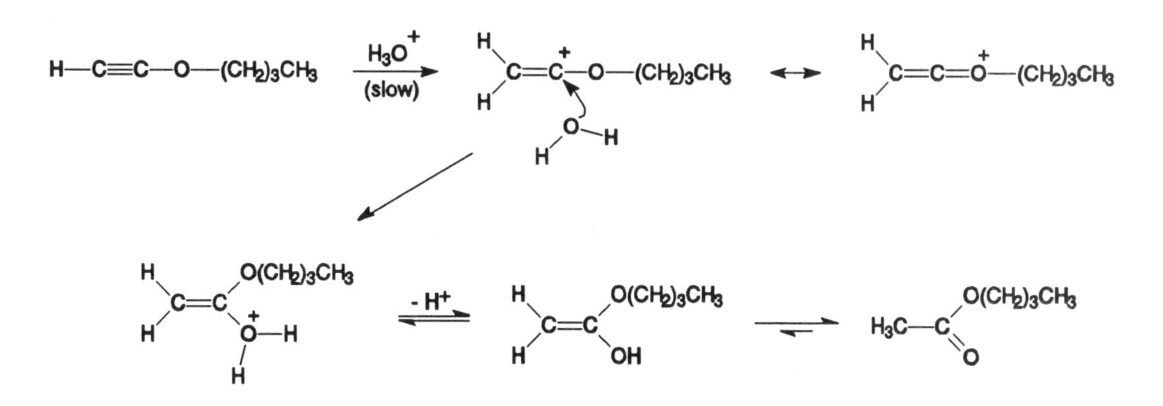

17. Halpern, J.; Tinker, H. B. *J. Am. Chem. Soc.* **1967**, *89*, 6427.

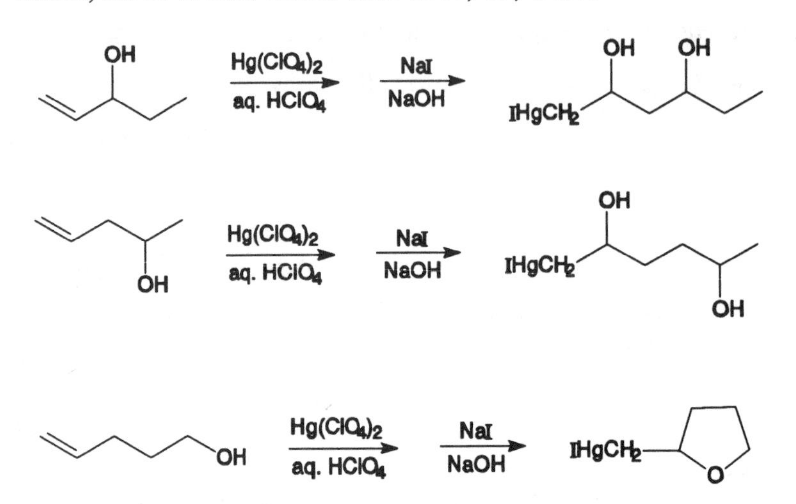

In general, the rate constants for oxymercuration correlate with Taft's σ* values. However, 1-penten-5-ol reacts by a pathway involving anchimeric assistance, so the rate is enhanced. The products were isolated as the organomercury iodides.

18. Kabalka, G. W.; Newton, Jr., R. J.; Jacobus, J. *J. Org. Chem.* **1978**, *43*, 1567.
Syn addition leads to the formation of the threo product from the (*E*) alkene and the erythro product from the (*Z*) alkene.

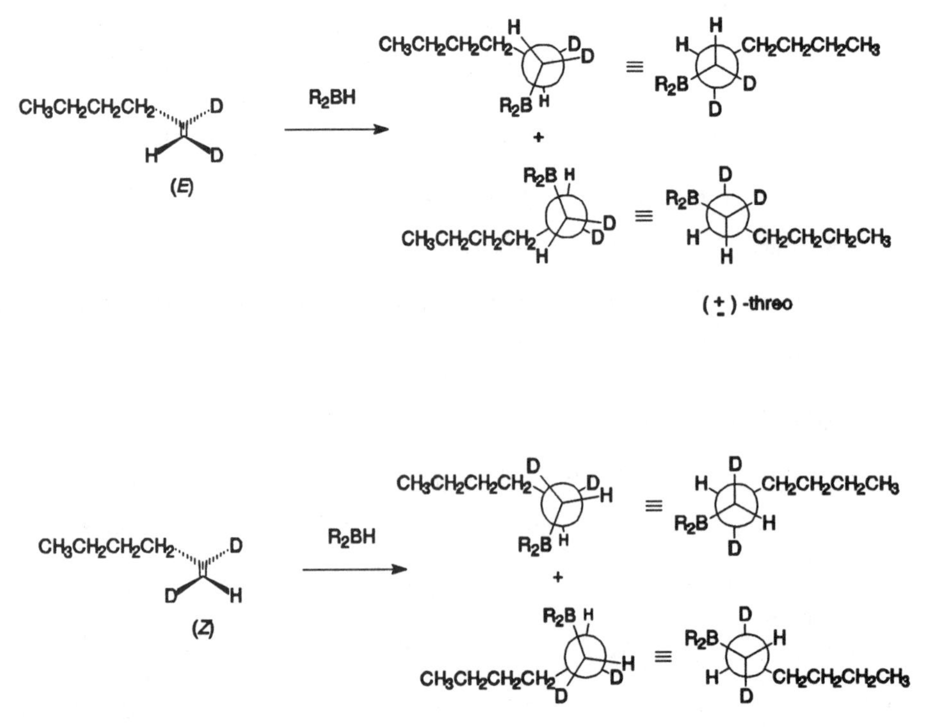

19. a) Brown, H. C.; Geoghegan, Jr., P. *J. Am. Chem. Soc.* **1967**, *89*, 1522; b) Brown, H. C.; Hammar, W. J. *J. Am. Chem. Soc.* **1967**, *89*, 1524.

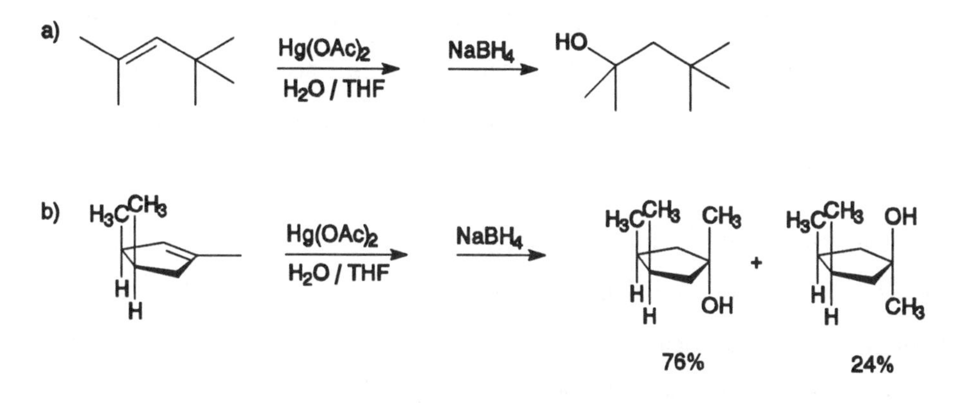

As indicated by the answer to part b), hydration occurs on the less sterically hindered side of the molecule.

20. Brown, H. C.; Kurek, J. T. *J. Am. Chem. Soc.* **1969**, *91*, 5647.

a) The product is *N*-cyclohexylacetamide.

b) The mechanism involves nucleophilic addition of CH_3CN to the mercurinium ion.

c) $Hg(NO_3)_2$ is used because nitrate is less nucleophilic than acetate or trifluoracetate, so it does not compete with solvent for nucleophilic addition to the mercurinium ion.

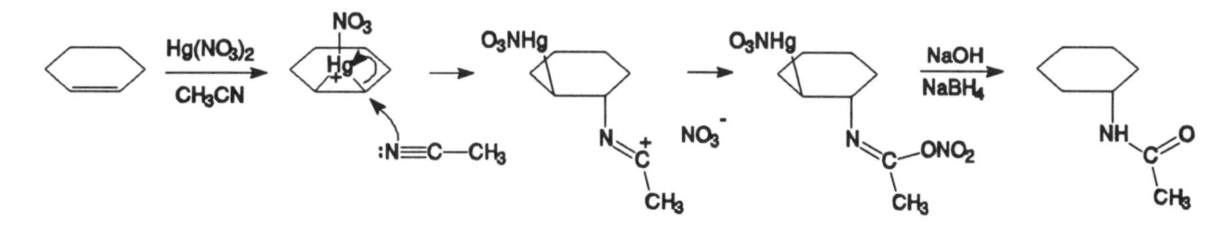

21. Swern, D. *J. Am. Chem. Soc.* **1947**, *69*, 1692.

In each case, the double bond with the methyl substituent or without the electron withdrawing group is more reactive.

22. Modro, A.; Schmid, G. H.; Yates, K. *J. Org. Chem.* **1977**, *42*, 3673.

The formation of the intermediate bromonium ion - bromide ion pair is facilitated by a polar solvent. In a less polar solvent, stabilization by electron-donating alkyl groups is more noticeable because the stabilization by solvent is less effective.

23. Ruasse, M.-F.; Motallebi, S.; Galland, B. *J. Am. Chem. Soc.* **1991**, *113*, 3440.

In both cases a neutral organic substrate is converted to a positively charged intermediate that adds a nucleophile to form the final product. Electron-releasing substituents accelerate the rates of both reactions. Both the departure of the leaving group in solvolysis and the departure of Br^-

are facilitated by electrophilic solvent (or by Br_2 in nonpolar solvent). Just as a carbocation - carbanion pair may return to the starting material in solvolysis, there is evidence that a bromonium - bromide (or tribromide) ion pair may return to starting materials in the electrophilic bromination of alkenes. The authors of the paper cited here also consider the possibility of a range of electrophilic bromination pathways, analogous to the range of substitution pathways ranging from S_N1 to S_N2.

24. There is no literature reference for this problem.

It may be argued that — for some purposes, at least — overly simplified mechanistic models may allow synthetic chemists to focus on synthetic strategy and not on detailed representations of reactive intermediates. Ultimately, however, representations of mechanisms and reactive intermediates that correspond to models developed in mechanistic studies may be the most useful in designing novel synthetic reactions.

Chapter 10

1. Dhar, M. L.; Hughes, E. D.; Ingold, C. K.; Mandour, A. M. M.; Maw, G. A.; Woolf, L. I. *J. Chem. Soc.* **1948**, 2093.

   For discussions of these results, see pp. 2107 and 2101, respectively, of the paper cited. The first two reactions show Saytzeff orientation, while the second two show Hofmann orientation. In each case, however, the additional methyl gives a greater statistical basis for 1-alkene formation.

2. Cristol, S. J.; Hause, N. L. *J. Am. Chem. Soc.* **1952**, *74*, 2193.

   H and Cl lie in the same plane in **9** but in **8** they do not.

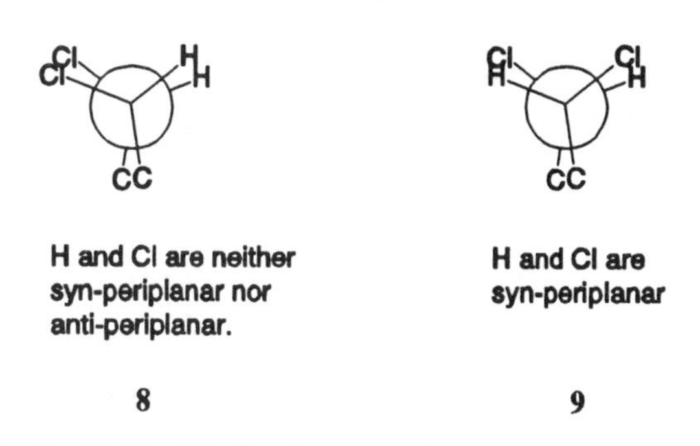

3. Kibby, C. L.; Lande, S. S.; Hall, W. K. *J. Am. Chem. Soc.* **1972**, *94*, 214.

   Syn elimination from *threo*-2-butanol-*3*-d_1 would give labeled *cis*-2-butene and unlabeled *trans*-2-butene. Anti elimination from *threo*-2-butanol-*3*-d_1 would give labeled *trans*-2-butene and unlabeled *cis*-2-butene. Syn elimination from *erythro*-2-butanol-*3*-d_1 would give labeled *trans*-2-butene and unlabeled *cis*-2-butene. Anti elimination from *erythro*-2-butanol-*3*-d_1 would give labeled *cis*-2-butene and unlabeled *trans*-2-butene. The reactions are illustrated for the threo diastereomer in the following figure.

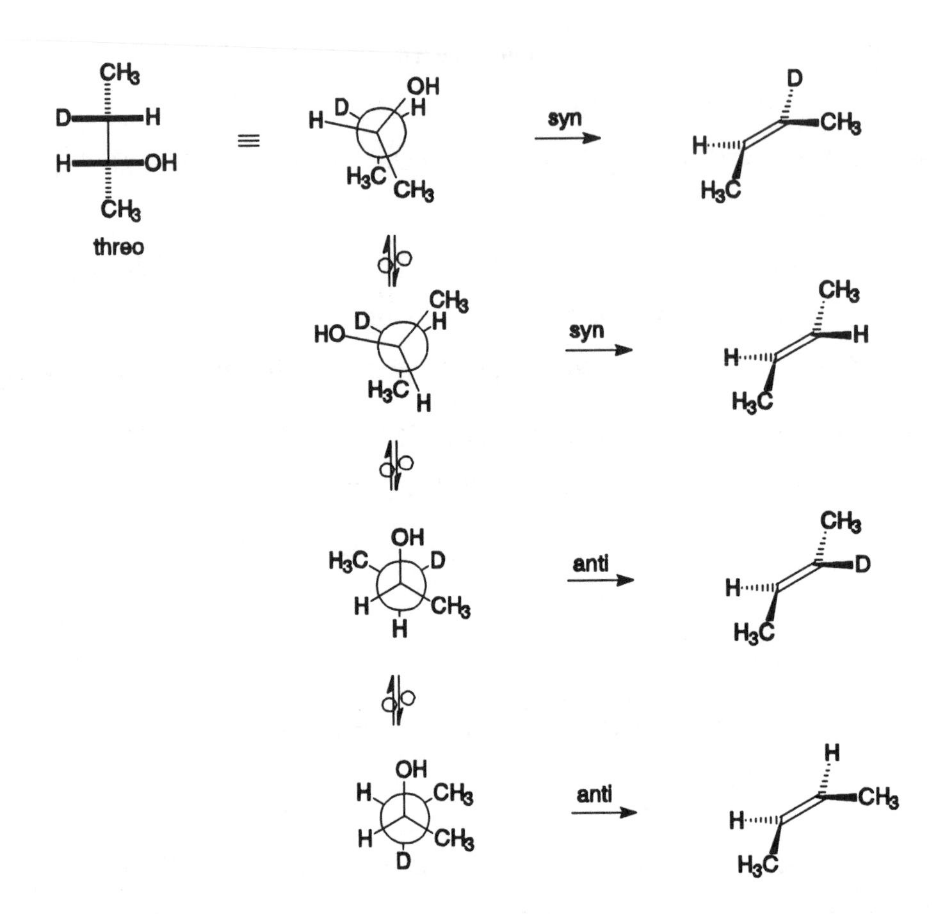

4. Searles, S.; Gortatowski, M. J. *J. Am. Chem. Soc.* **1953**, *75*, 3030.

As formulated by the researchers, the reaction occurs by a two-step mechanism in which the rate-limiting step is fragmentation of an alkoxide ion.

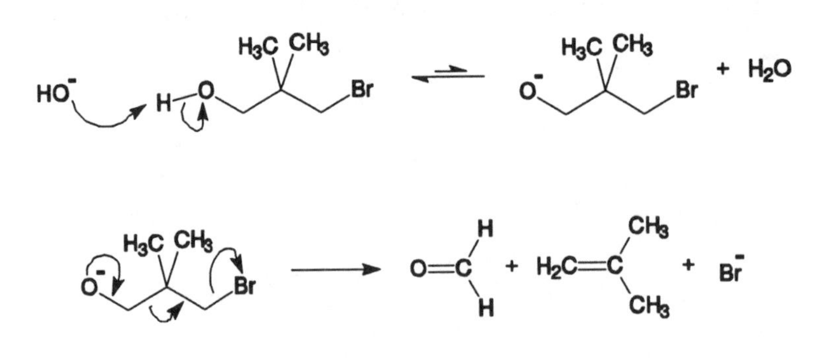

5. Grovenstein, Jr., E.; Lee, D. E. *J. Am. Chem. Soc.* **1953**, *75*, 2639; Cristol, S. J.; Norris, W. P. *J. Am. Chem. Soc.* **1953**, *75*, 2645.

The elimination in acetone is a concerted anti elimination from the carboxylate ion.

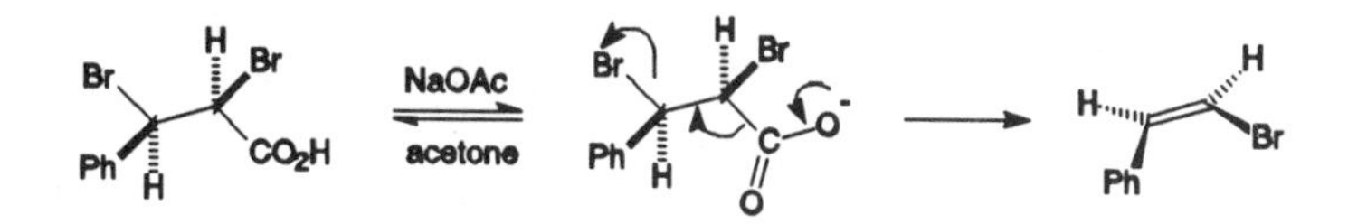

In a more polar solvent, ionization is competitive with elimination. The resulting carbocation can undergo conformational change before elimination of CO_2, so the more stable product is obtained.

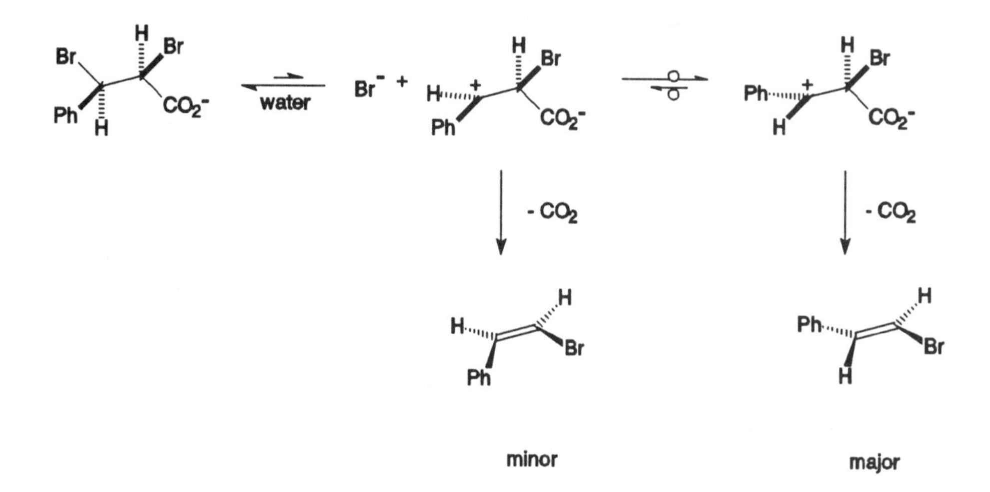

6. Ölwegård, M.; Ahlberg, P. *J. Chem. Soc., Chem. Commun.* **1989**, 1279; **1990**, 788.

50 and **51** should give different distributions of *deuterium-containing* (*E*)- and (*Z*)-ethylidene-indene, thus allowing determination of the relative rate constants for syn and anti elimination.

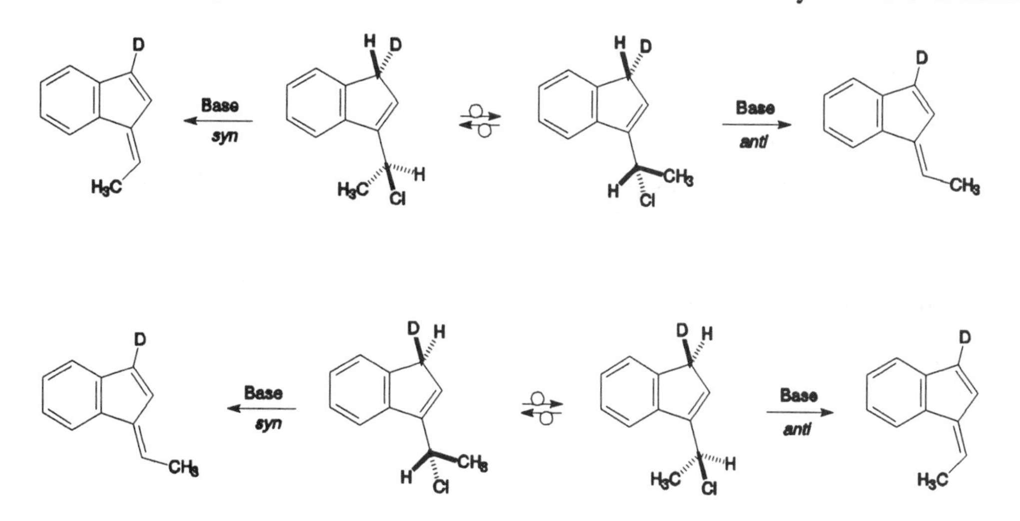

7. Kurtz, R. R.; Houser, D. J. *J. Org. Chem.* **1981**, *46*, 202.

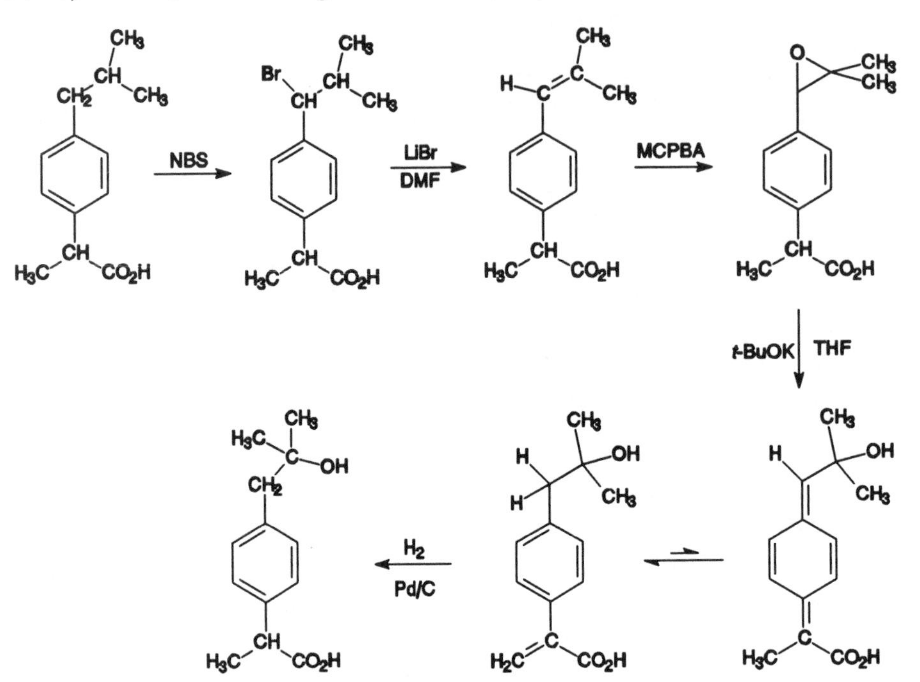

8. Goering, H. L.; Espy, H. H. *J. Am. Chem. Soc.* **1956**, *78*, 1454.

a) anti periplanar elimination is preferred; products of two are 1-chlorocyclohexene; the third goes to 1,3-cyclohexadiene.

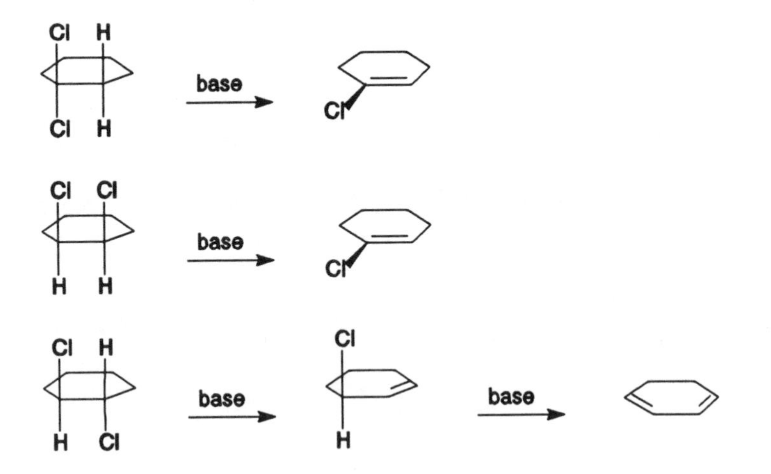

b) Presence of a Cl substituent makes the H on that same carbon atom more acidic, so that is the preferred route of elimination.

9. Nguyen Trong Anh, *Chem. Commun.* **1968**, 1089.

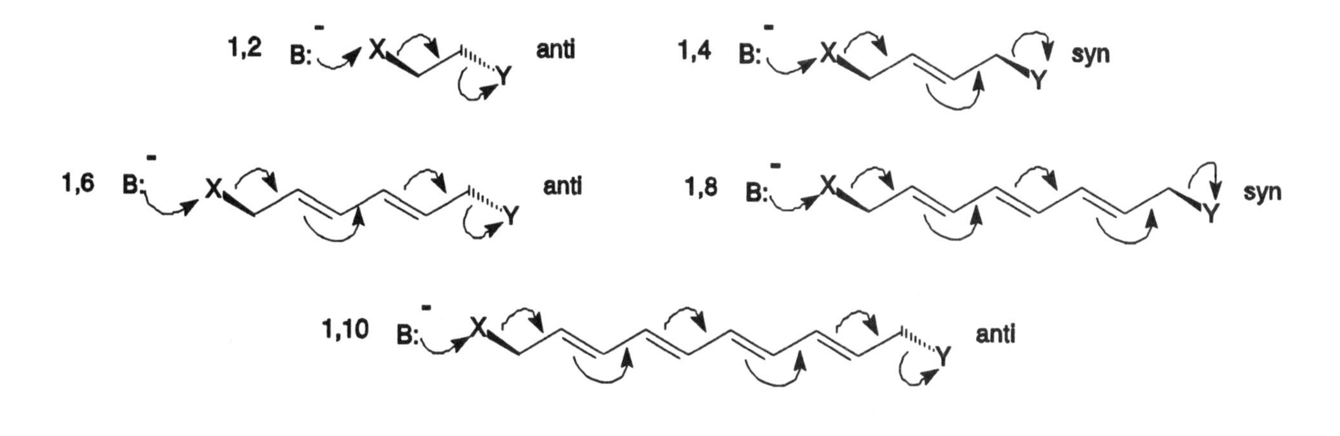

10. Cope, A. C.; LeBel, N. A.; Lee, H.-H.; Moore, W. R. *J. Am. Chem. Soc.* **1957**, *79*, 4720. There are differing steric requirements for the transition structures for the two reactions. The two transition structures for the Cope elimination, shown here

represent eclipsed conformations of the alkyl branch on which elimination occurs. Any lowering of the transition structure energy due to greater stabilization of the incipient alkene appears to be offset by the steric hindrance of an eclipsed methyl-hydrogen interaction. In addition, Cope and coworkers proposed that the transition structure for formation of ethene in equation 10.70 should be lower than that for formation of propene because the conformation required for anti elimination of propene has an unfavorable methyl - trialkylammonium interaction.

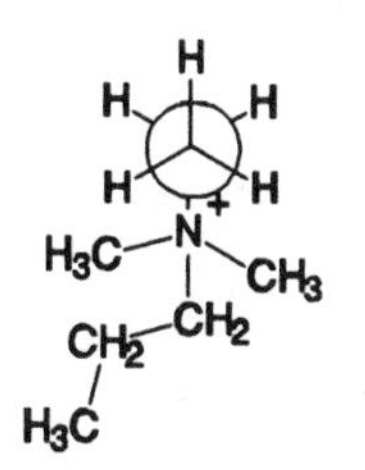

Conformation for elimination of ethene

unfavorable gauche interaction

Conformation for elimination of propene

11. Gandini, A.; Plesch, P. H. *J. Chem. Soc.* **1965**, 6019.

The reaction was proposed to be a concerted elimination that should exhibit syn stereochemistry.

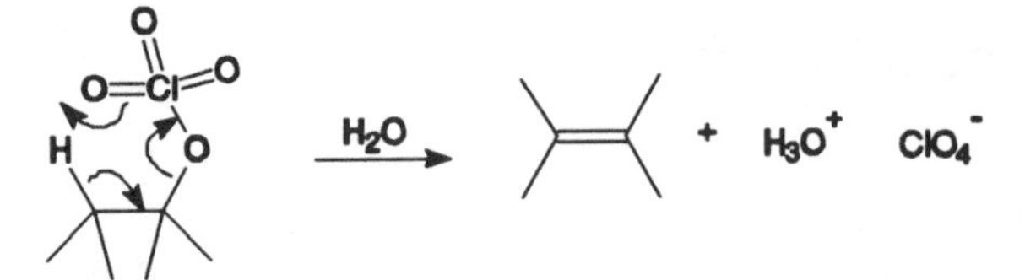

12. O'Connor, G. L.; Nace, H. R. *J. Am. Chem. Soc.* **1953**, *75*, 2118.

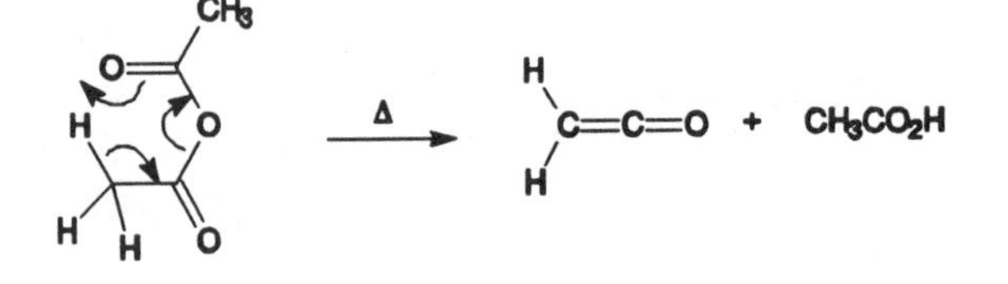

13. Knözinger, H. in *The Chemistry of the Hydroxyl Group*, Part 2; Patai, S., ed.; Wiley-Interscience: London, 1971; pp. 660-661 and references therein.

The alkoxide undergoes a fragmentation reaction.

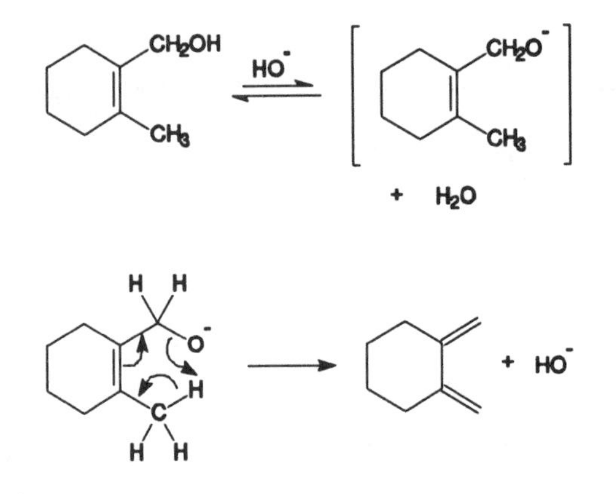

It is interesting that the literature mecahnism proposes that the electrons in the C—O bond (and not one of the three electron pairs localized on the O^-) should abstract the proton from the methyl group in a six electron cyclic mechanism that suggests an aromatic transition state. An alternative mechanism in which an electron pair on O^- abstracts the proton and the electrons in the C—O bond leave with the OH group in a concerted process would be an 8-electron cyclic transition state. Unless there is some orthogonality of electron pairs, this would involve an antiaromatic transition structure. Therefore, a nonconcerted pathway might be lower in energy. A possible nonconcerted pathway would be one involving prior hydrogen abstraction by the alkoxide ion, leaving a CH_2^- group that can then undergo electron reorganization with expulsion of hydroxide ion. However, a very rough estimate of the difference in pKa's of an alkane and an alcohol is about 25, so the equilibrium necessary for the nonconcerted mechanism would be very unfavorable.

14. Acharya, S. P.; Brown, H. C. *Chem. Commun.* **1968**, 305.

The base should be as bulky as possible. The investigators used $(CH_3CH_2)_3CO^-K^+$ in $(CH_3CH_2)_3COH$ solution.

15. Barton, D. H. R.; Rosenfelder, W. J. *J. Chem. Soc.* **1951**, 1048.

In 2ß,3α-dibromocholestane both bromine atoms are axial, so the anti-periplanar orientation for E2 elimination is feasible. In 3ß,4α-dibromocholestane, however, the two bromine atoms are equatorial, so an anti-periplanar arrangement is not feasible.

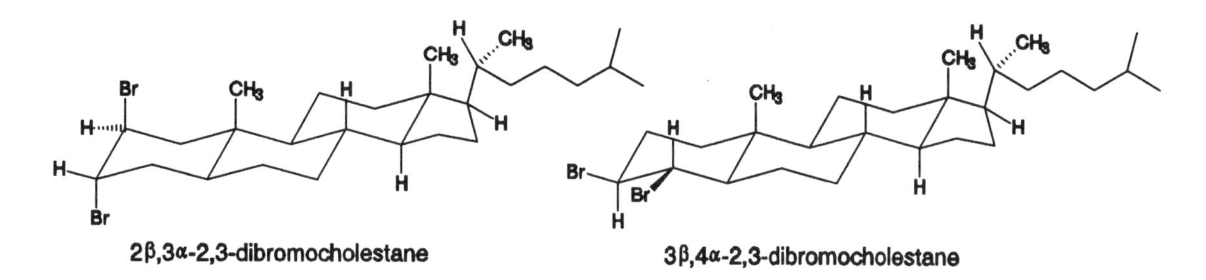

16. Kashelikar, D. V.; Fanta, P. E. *J. Am. Chem. Soc.* **1960**, *82*, 4930.

The mechanism is proposed to be a concerted reaction like that of the Chugaev or Cope reactions.

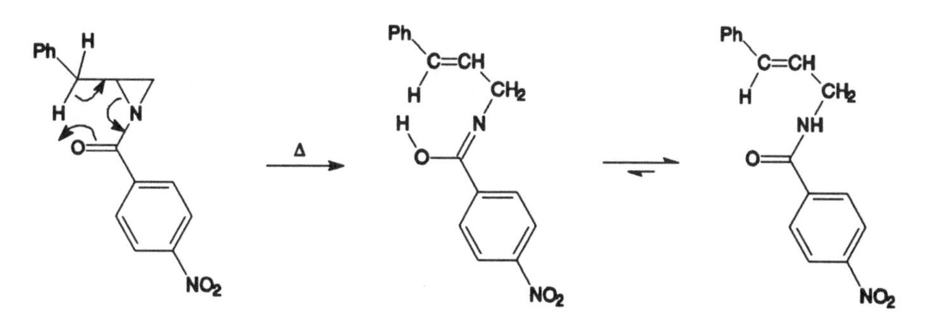

17. Noyce, D. S.; Weingarten, H. I. *J. Am. Chem. Soc.* **1957**, *79*, 3093.

In the trans isomer interaction of the methoxy group with the acid chloride function is not feasible, so the acid chloride is isolated.

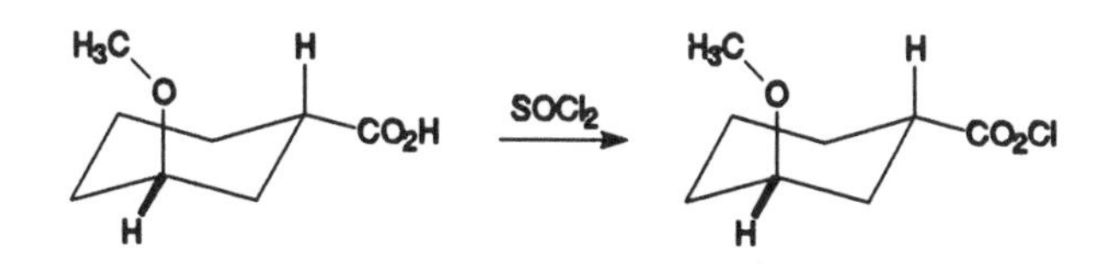

With the cis isomer, the methoxy group reacts with the acid chloride to yield the reported products as a result of S_N2 or E2 attack by halide or other anion.

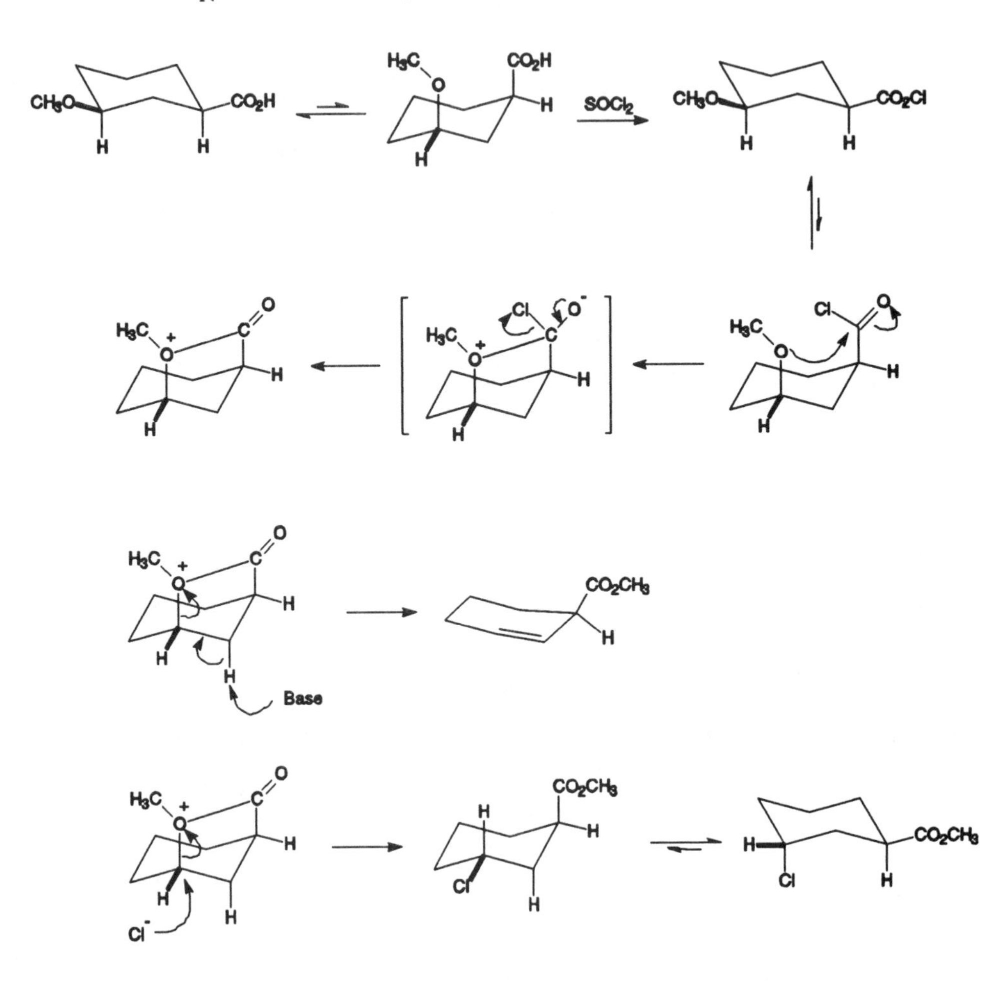

18. a. Alexander, E. R.; Mudrak, A. *J. Am. Chem. Soc.* **1950**, *72*, 1810.

In the cis case, the hydrogen on the carbon atom with the phenyl group is not accessible in a six-membered transition structure. With the trans isomer, both hydrogens are accessible, although formation of the conjugated product is favored.

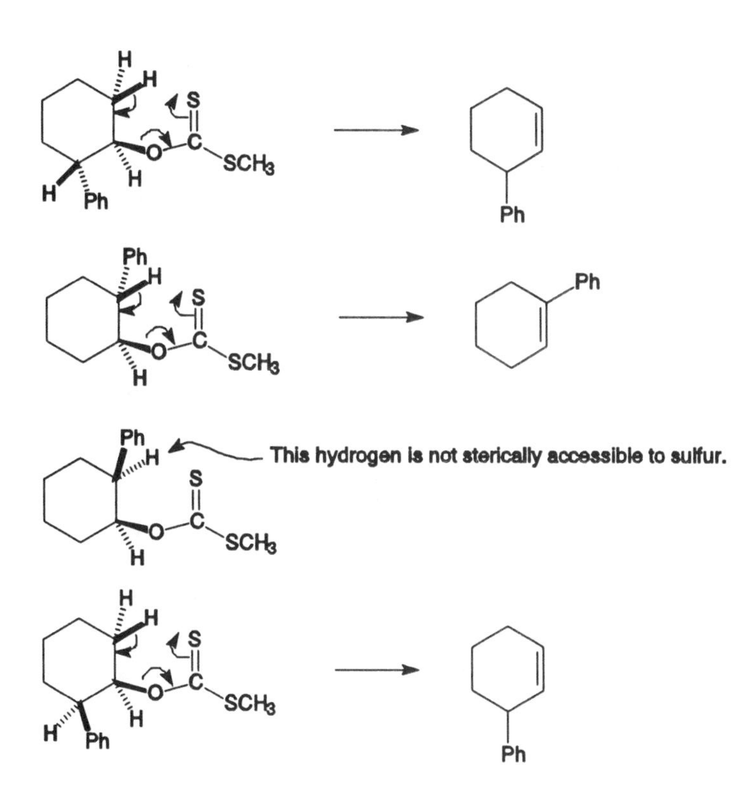

b. Alexander, E. R.; Mudrak, A. *J. Am. Chem. Soc.* **1950**, *72*, 3194.

The cis isomer has no sterically accesible hydrogen atom for a concerted reaction involving a six-membered transition structure.

c. Cope, A. C.; LeBel, N. A. *J. Am. Chem. Soc.* **1960**, *82*, 4656.

The five-membered (H—C—C—N^{+}—O^{-}) ring transition structure shown at the right is not sterically feasible in the smallest ring.

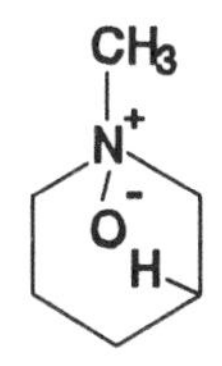

19. Kende, A. S. *Org. React.* **1960**, *11*, 261.

The first step is a 1,3-elimination reaction.

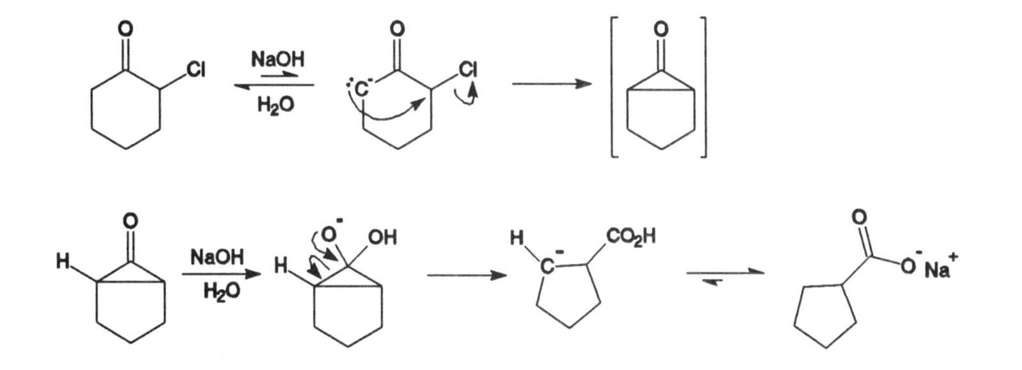

20. Curtin, D. Y.; Stolow, R. D.; Maya, W. *J. Am. Chem. Soc.* **1959**, *81*, 3330.

The trans isomer cannot have an axial trimethylammonium group (because it would also have to have an axial *t*-butyl group), so it cannot undergo an E2 reaction. However, reaction of the substrate (at a methyl group) with *t*-butoxide can produce an S_N2 reaction.

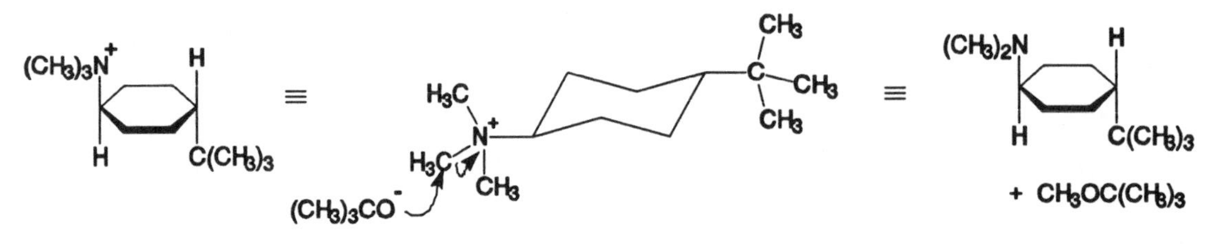

The cis isomer can undergo both substitution and elimination.

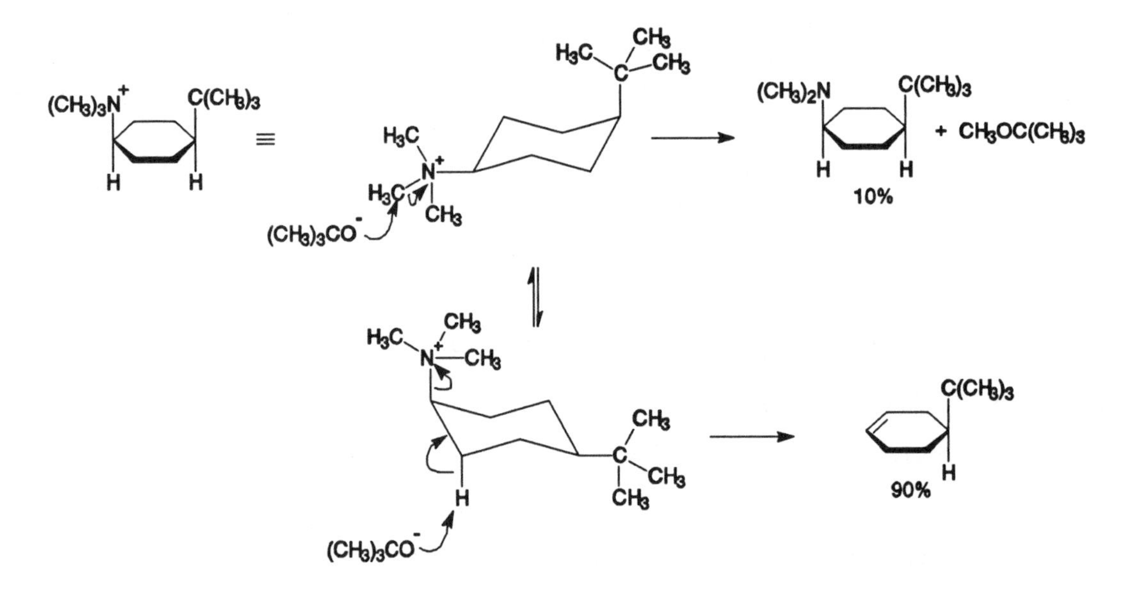

21. Saunders, Jr., W. H.; Cockerill, A. F. *Mechanisms of Elimination Reactions*; Wiley-Interscience: New York, 1973; p. 173.

The former is less substituted, but the latter has *t*-butyl - methyl repulsion. If Saytzeff means giving the more stable alkene, then it is Saytzeff orientation. If Saytzeff means giving the more highly substituted alkene, then it is Hofmann orientation.

Chapter 11

1. There is no literature reference for this problem.

   For the thermal reaction, the HOMO of octatetraene is ψ_4. There are three nodes in ψ_4, so the sign of the coefficient for ϕ_1 has the opposite sign from that of ϕ_8. Therefore, the thermal reaction should be conrotatory. In the exicted state, HOMO is ψ_5. Now the sign of the coefficient for ϕ_1 is the same as that for ϕ_8, so the photochemical reaction should be disrotatory.

2. There is no literature reference for this problem.

   The thermal [1,7] hydrogen shift is analyzed in terms of a transition structure resembling a hydrogen atom and a heptatrienyl radical. For the heptatrienyl radical, HOMO is ψ_4, which has three nodes. The coefficient for ϕ_1 has the opposite sign from that of ϕ_7. Therefore, in order to maintain a bonding relationship from H to C1 while a bond from H to C7 is forming, the reaction would have to be antarafacial with respect to the heptatrienyl radical.

   The thermal [1,9] hydrogen shift is analyzed in terms of a transition structure resembling a hydrogen atom and a nonatetraenyl radical. For the nonatetraenyl radical, HOMO is ψ_5. There are four nodes in ψ_5, so the coefficients of ϕ_1 and ϕ_9 have the same sign. Therefore, a bonding relationship between H and C1 can be maintained while a bonding relationship between H and C9 is being established, so the suprafacial [1,9] hydrogen shift is allowed by the principles of orbital symmetry.

3. There is no literature reference for this problem.

The suprafacial-suprafacial $[_{\pi}2_s + _{\pi}6_s]$ cycloaddition is analyzed in terms of a transition structure with a plane of symmetry (bisecting the C3-C4 bond of hexatriene and the C1-C2 bond of ethene) and is seen to be forbidden by the principles of orbital symmetry.

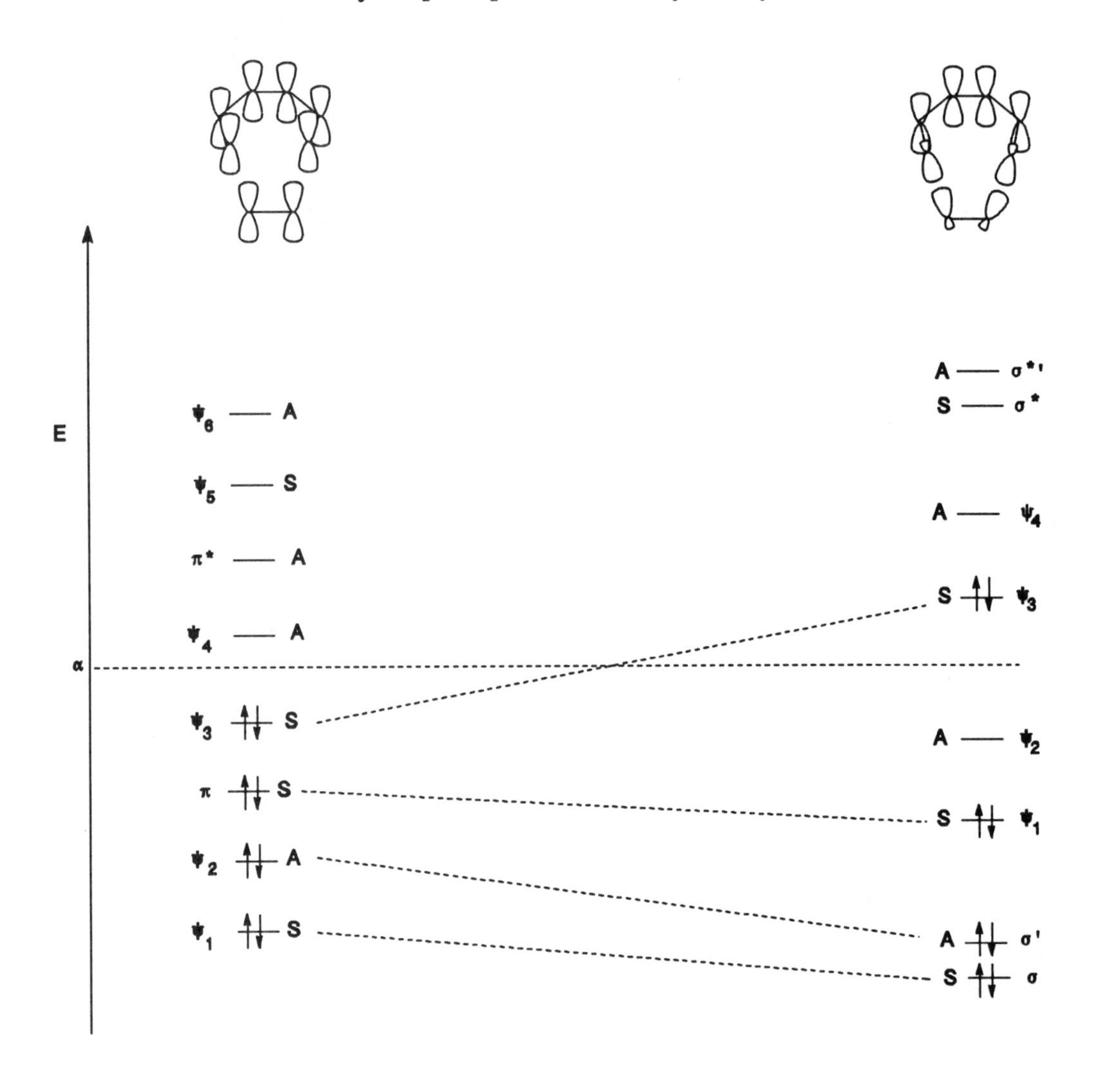

The suprafacial-suprafacial $[_{\pi}4_s + _{\pi}4_s]$ cycloaddition is analyzed in terms of a transition structure with a plane of symmetry (bisecting the C2-C3 bond in each butadiene molecule) and is seen to be forbidden by the principles of orbital symmetry.

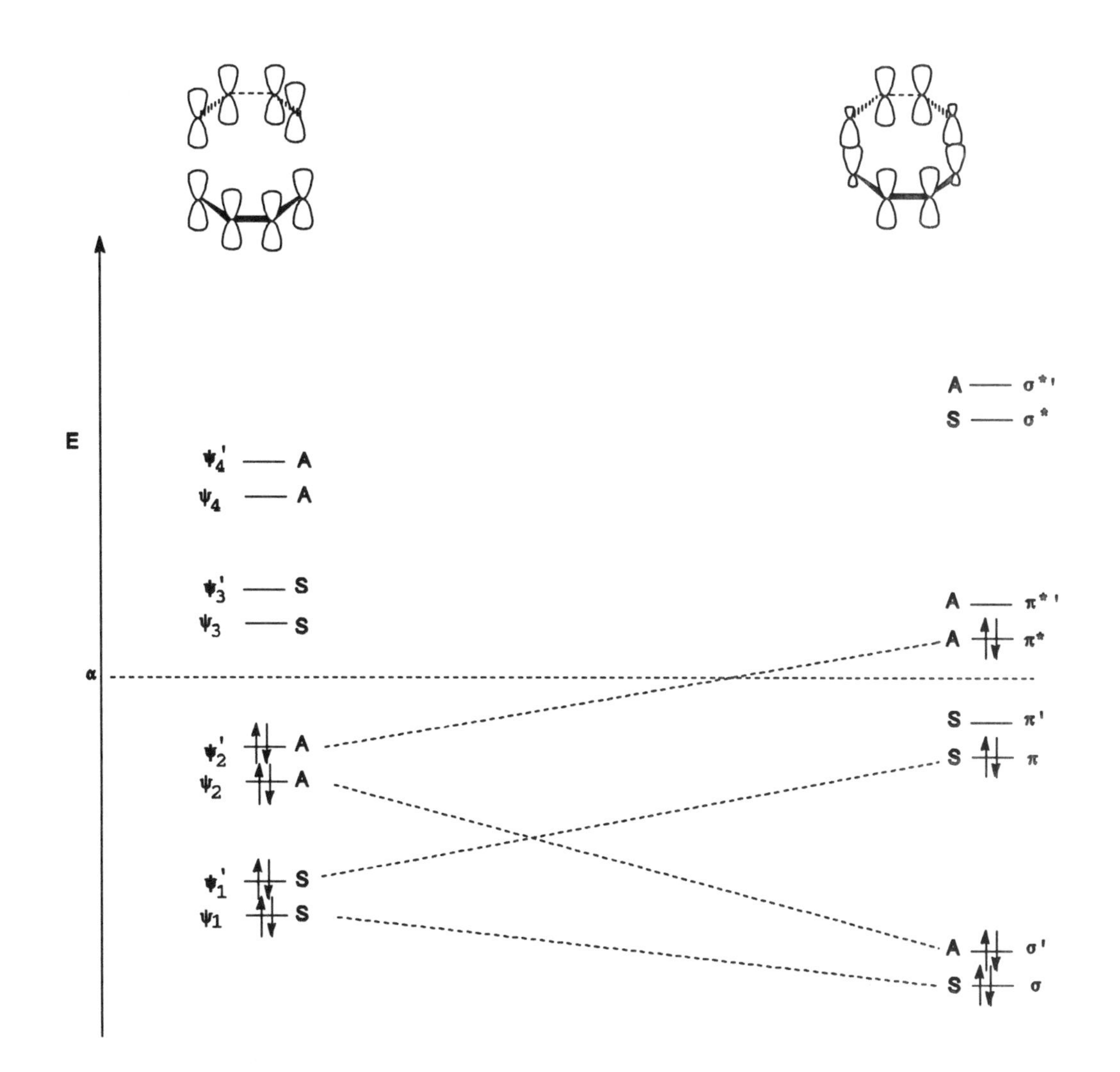

4. There is no literature reference for this problem.

a. The antarafacial-antarafacial transition structure is shown below:

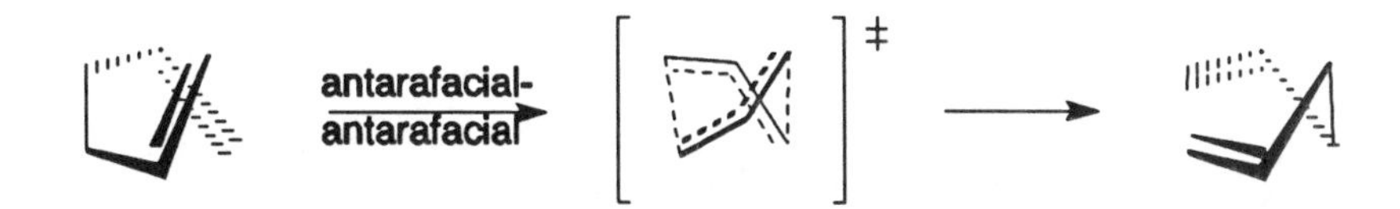

The problem is analyzed by considering ψ_2 of one allyl fragment (drawn at the back of the transition structure) and ψ_3 of the other allyl fragment (drawn here at the front of the transition structure). It may be seen that an antibonding relationship exists between the orbitals of the incipient C1-C6 bond, so the reaction is forbidden by the principles of orbital symmetry.

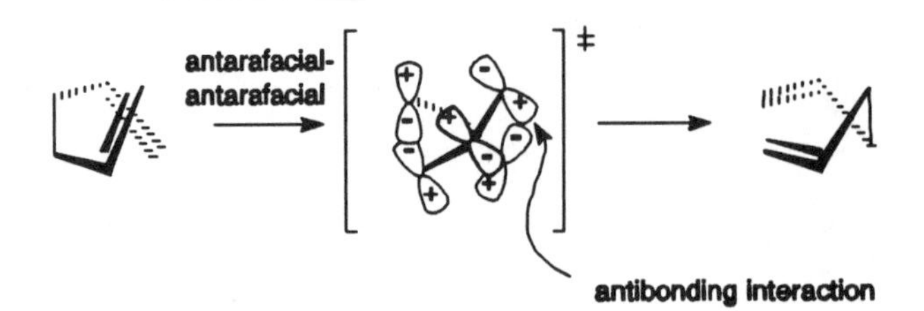

b. ES-1 of the reactants does not correlate with ES-1 of the product. Instead, it correlates with a much higher energy excited state of the product. Therefore, there would be an avoided crossing (and an electronic barrier) for conversion of ES-1 of the reactants into ES-1 of the product by the suprafacial-suprafacial pathway.

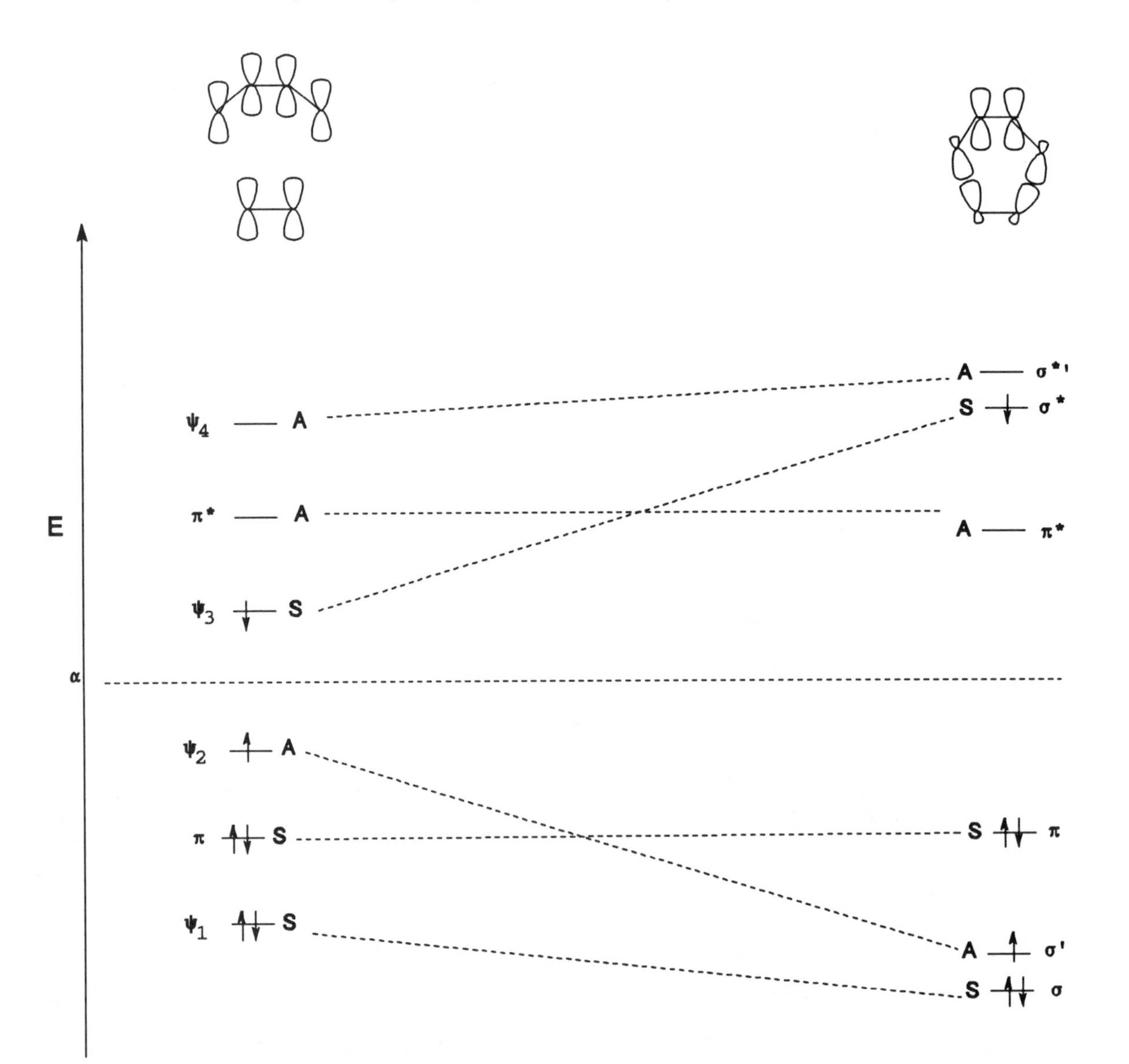

5. Longuet-Higgins, H. C.; Abrahamson, E. W. *J. Am. Chem. Soc.* **1965**, *87*, 2045. A discussion of the construction of orbital correlation diagrams using HMO theory has been given by Dalton, J. C.; Friedrich, L. E. *J. Chem. Educ.* **1975**, *52*, 721.

For the cation, only σ is populated. In the disrotatory reaction a plane of symmetry is maintained, so a symmetric σ orbital correlates with ψ_1 of the allyl cation and the reaction is allowed. For the conrotatory pathway, a C_2 rotation axis is maintained. Now the σ orbital is symmetric, but ψ_1 of allyl is antisymmetric. Therefore the σ orbital correlates with ψ_2 of allyl, so the reaction is thermally forbidden.

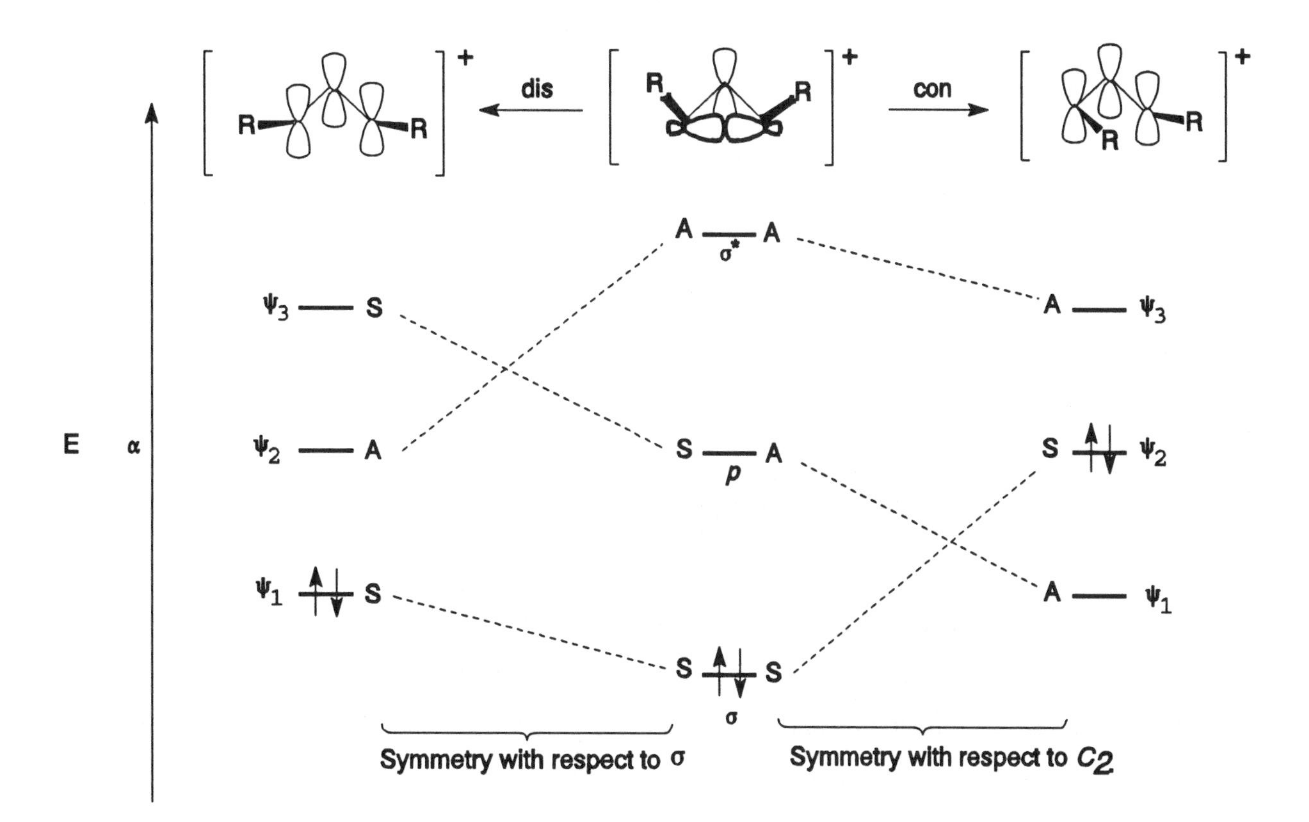

In the anion, both σ and p of the cyclopropyl moiety are populated. Orbital symmetries are designated as above. Now the cyclopropyl anion correlates with the allyl anion by the conrotatory pathway.

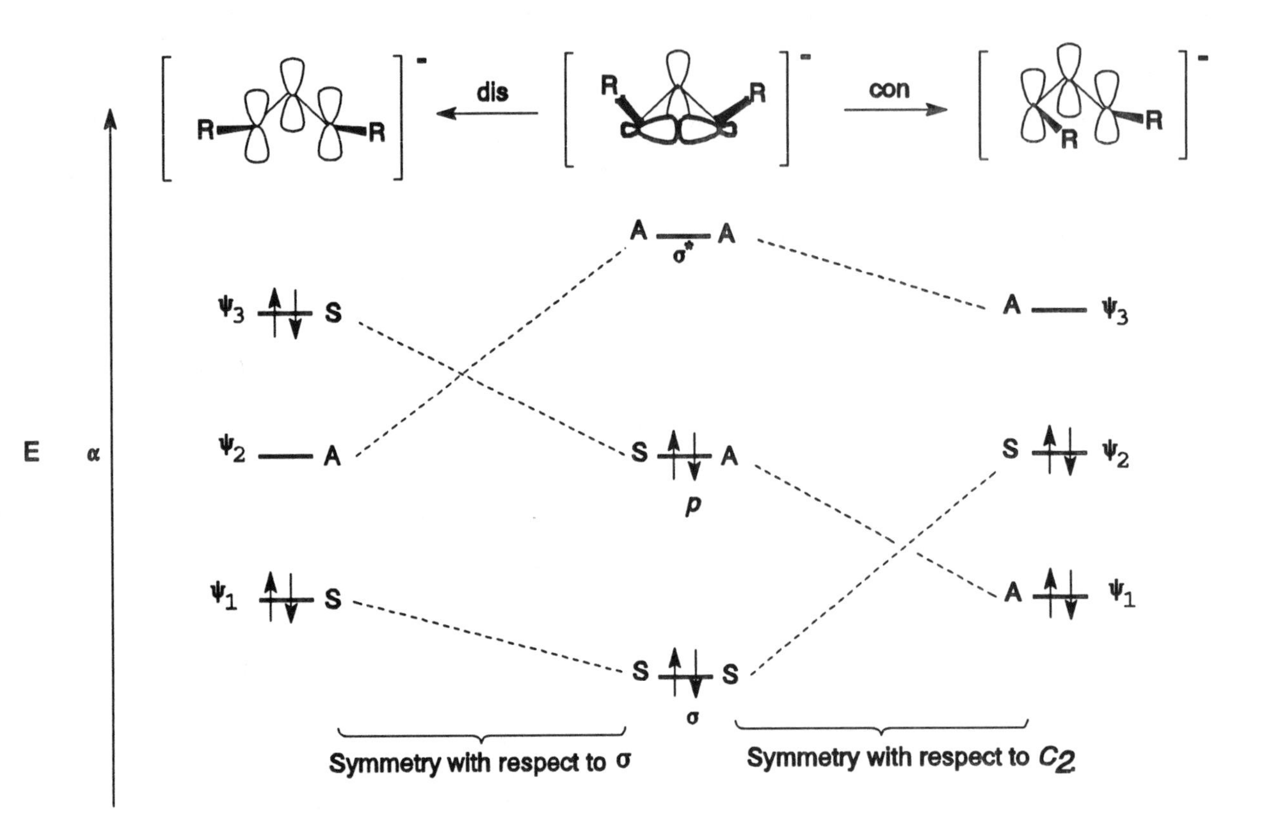

The state correlation diagrams are constructed from the molecular orbital correlation diagrams. For the cation, it is evident that the ground state (GS) of the cyclopropyl cation correlates with the ground state of the allyl cation for the disrotatory opening. (The state symmetry designations are the products of the MO symmetry designations for the molecular orbitals populated in each state.) However, the cyclopropyl cation correlates with the allyl cation for a photochemical reaction, because ES-1 of cyclopropyl correlates with ES-1 of allyl for that process.

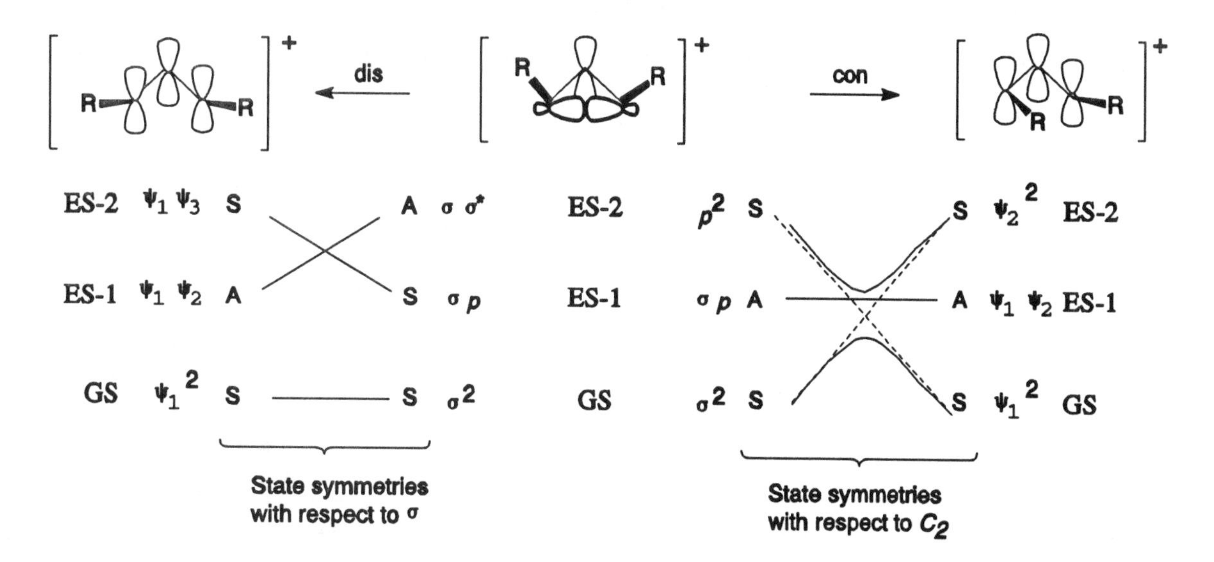

The ground state of the cyclopropyl anion correlates with the ground state of the allyl anion by the conrotatory pathway. However, the first excited state of the cyclopropyl anion correlates with the first excited state of the allyl anion by the disrotatory pathway.

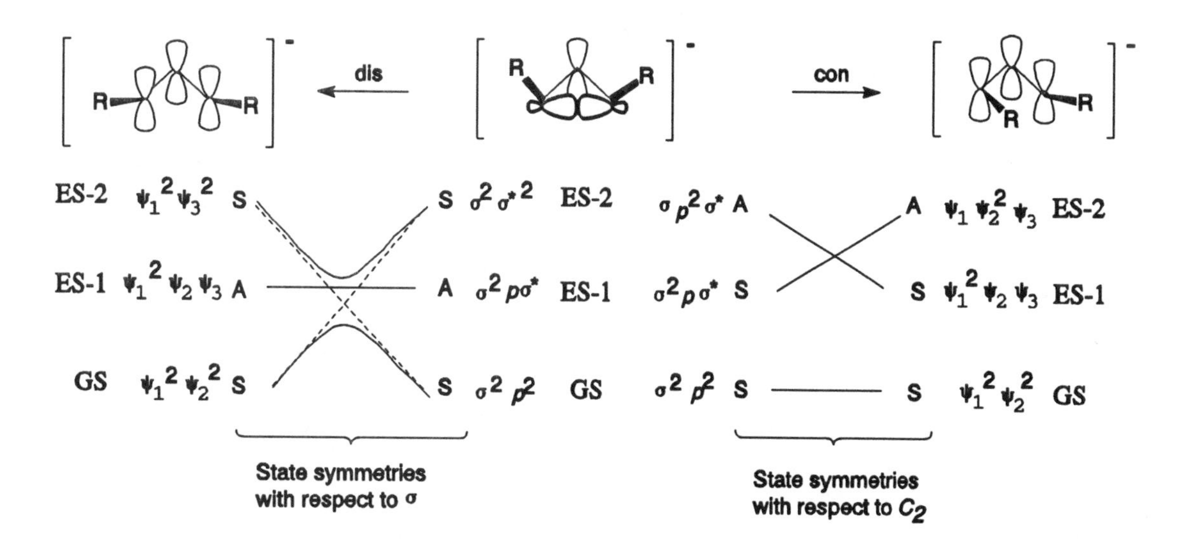

6. There is no literature reference for this problem.

a. The Möbius MO's for cyclopropenyl are calculated on the right below. By comparison, the determinant for Hückel cyclopropenyl is shown on the left.

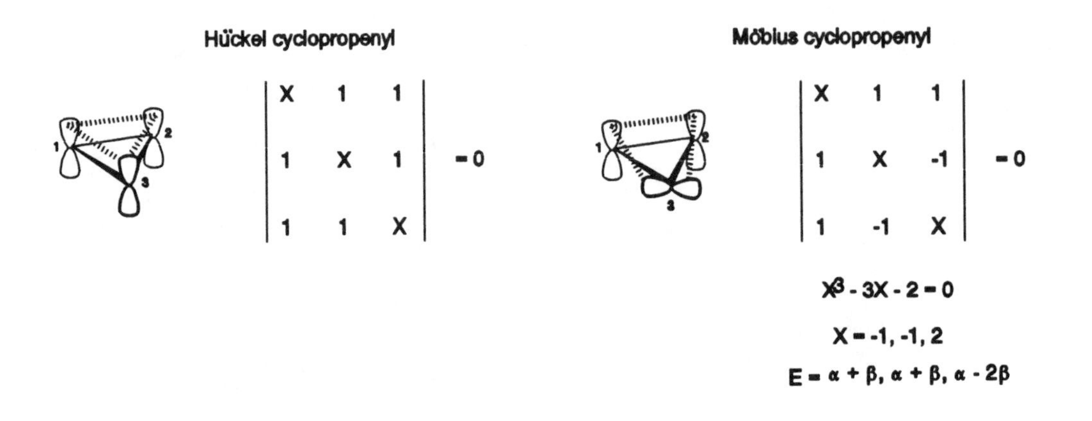

b. For the cation:

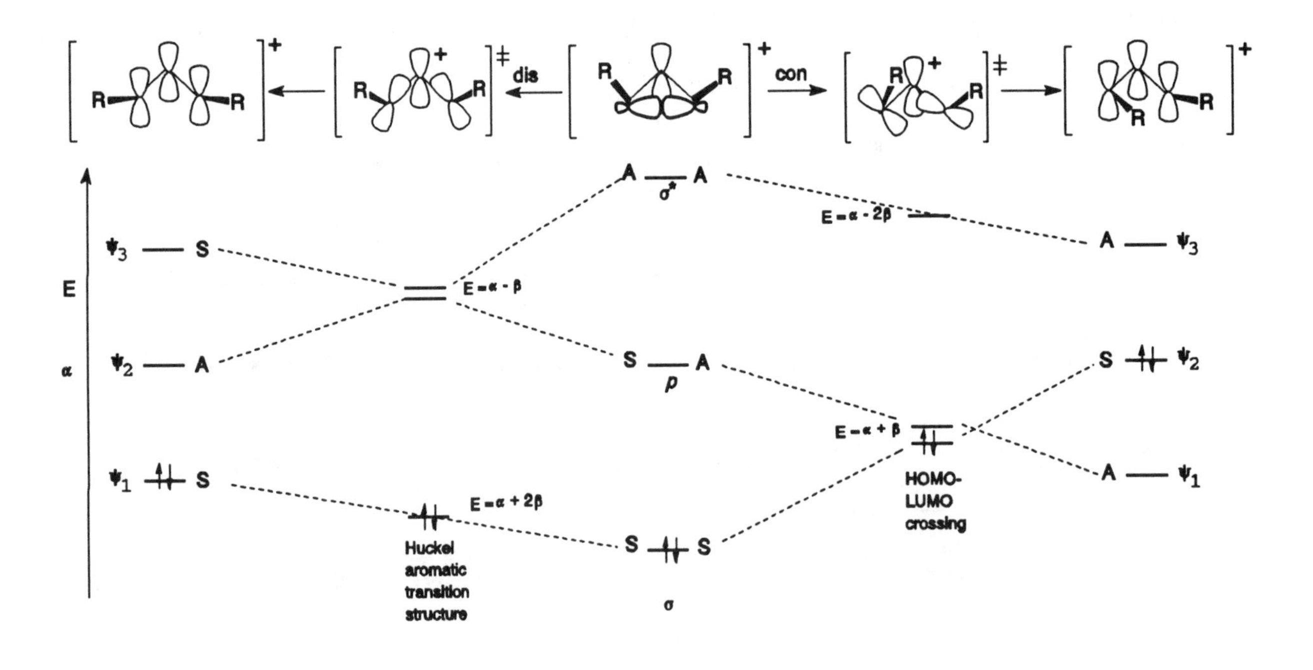

For the anion:

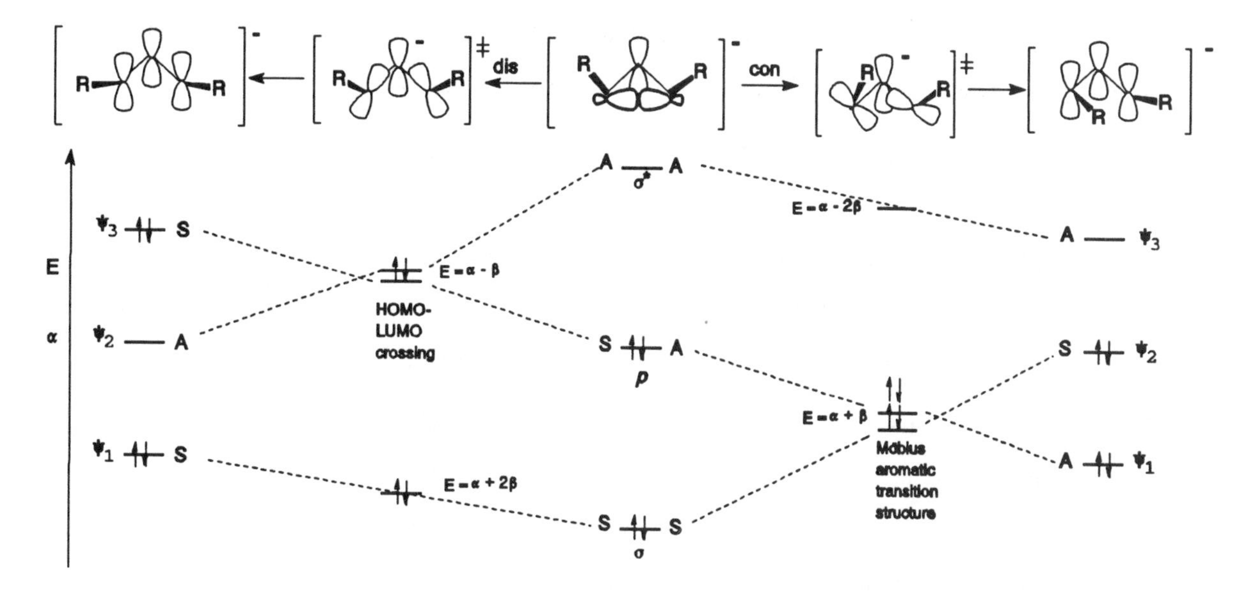

7. a. The pathway shown is a $[_{\pi}4_{a}]$ process. The number of $(4n+2)_s$ components is 0, the number of $(4r)_a$ components is 1, the total is odd, and the reaction is allowed. The transition stucture should be Möbius aromatic.

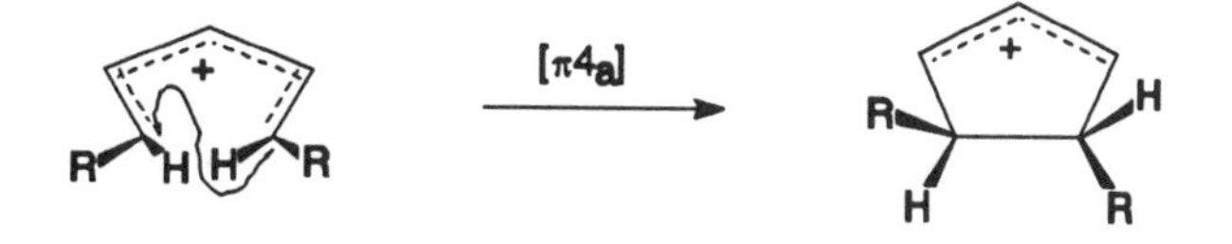

b. The pathway shown is a $[_{\pi}6_{s}]$ process. The number of $(4n+2)_s$ components is 1, the number of $(4r)_a$ components is 0, the total is odd, and the reaction is allowed. The transition stucture should be Hückel aromatic.

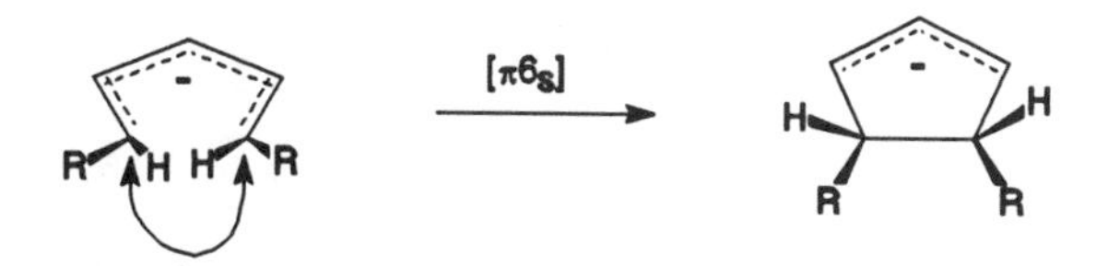

c. The pathway shown is a $[_{\pi}6_{s} = {_{\pi}4_{s}}]$ process. The number of $(4n+2)_s$ components is 1, the number of $(4r)_a$ components is 0, the total is odd, and the reaction is allowed. The transition stucture should be Hückel aromatic.

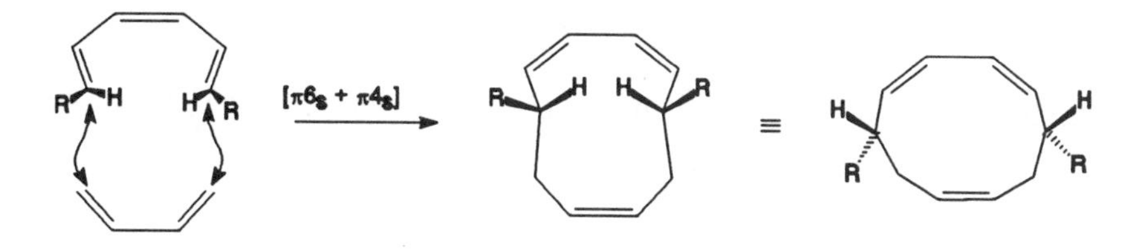

8. Fráter, G.; Schmid, H. *Helv. Chim. Acta* **1968**, *51*, 190.

   The authors of the study categorized the reaction as a [5,5] sigmatropic rearrangement, which is an allowed suprafacial-suprafacial process.

9. Vogel, E. *Liebigs Ann. Chem.* **1958**, *615*, 1; Also see Hammond, G. S.; DeBoer, C. D. *J. Am. Chem. Soc.* **1964**, *86*, 899.

   The cis isomer can achieve the conformation shown for a rapid Cope rearrangement. In the trans isomer, the two double bonds are not oriented properly for a concerted Cope rearrangement. It is likely that the reaction proceeds by homolytic cleavage of the cyclobutane σ bond.

10. Doering, W. von E.; Wiley, D. W. *Tetrahedron* **1960**, *11*, 183.

    The reaction shown in equation 11.24 is a $[_\pi 2_s + {}_\pi 8_s]$ cycloaddition, which is allowed by a suprafacial - suprafacial pathway. However, the reaction in equation 11.25 is a $[_\pi 2_s + {}_\pi 6_s]$ cycloaddition, which requires an SA or AS pathway.

11. a. Bates, R. B.; McCombs, D. A. *Tetrahedron. Lett.* **1969**, 977.

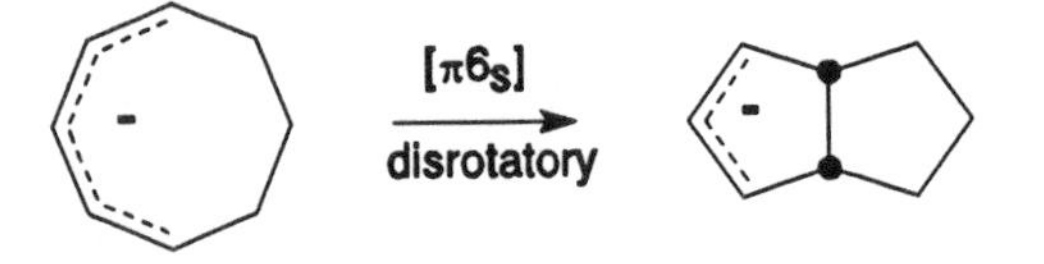

b. Pomerantz, M.; Wilke, R. N.; Gruber, G. W.; Roy, U. *J. Am. Chem. Soc.* **1972**, *94*, 2752.

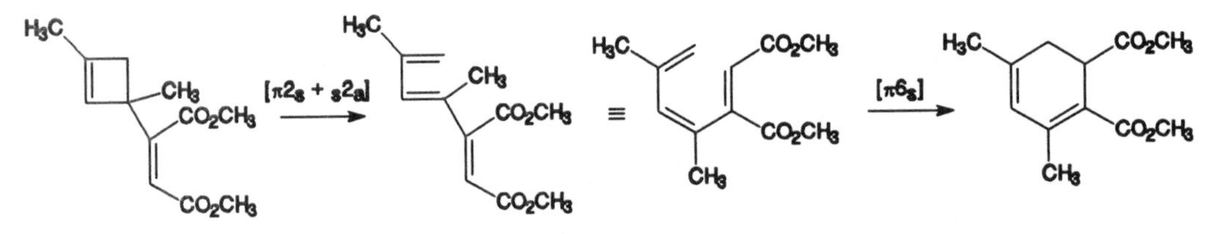

c. Arnold, B. J.; Sammes, P. G. *J. Chem. Soc., Chem. Commun.* **1972**, 1034; Arnold, B. J.; Sammes, P. G.; Wallace, T. W. *J. Chem. Soc., Perkin Trans. 1* **1974**, 415.

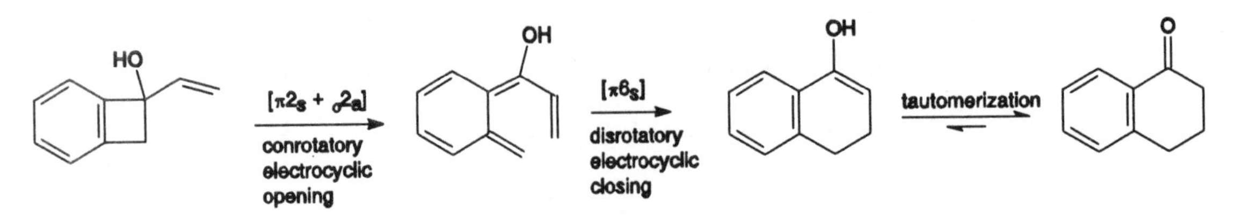

It is notable that the electrocyclic opening produces an intermediate with the oxygen in the (*E*) configuration, which facilitates closure to the enol of the product.

12. Brown, J. M. *Chem. Commun.* **1965**, 226.

The Wittig reaction produces *cis*-6-vinylbicyclo[3.1.0]hex-2-ene, which undergoes a rapid Cope rearrangement to give the product.

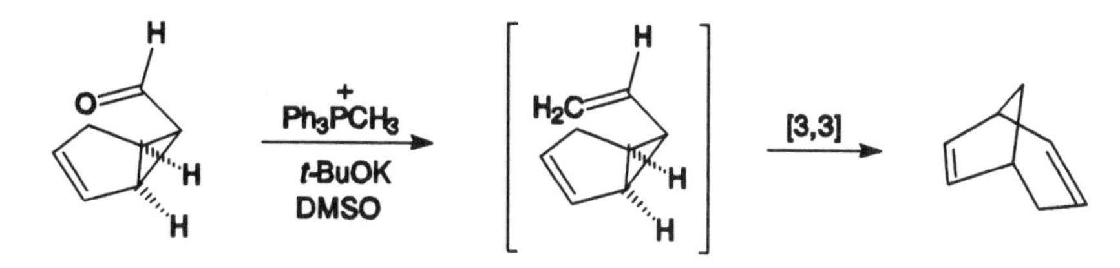

13. Curtin, D. Y.; Johnson, Jr., H. W. *J. Am. Chem. Soc.* **1956**, *78*, 2611.

The intermediate cyclohexadienone can undergo a retro-Claisen with migration of the methallyl group or can undergo a Cope rearrangement with either the allyl or methylallyl group.

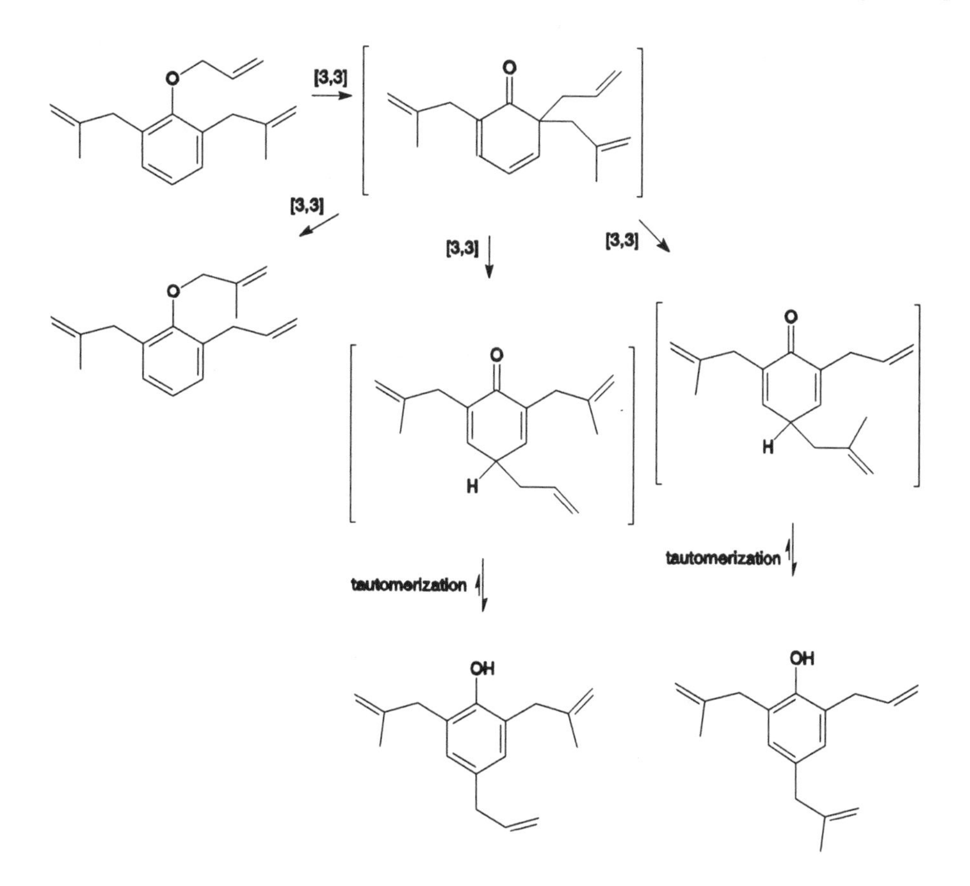

14. Dowd, P.; personal communication to Woodward and Hoffmann (reference 15), p. 143. Also, Fleming, I.; personal communication to Woodward and Hoffmann, p. 143.

The reaction is allowed, with $m = 4$ and $n = 2$.

15. Arnett, E. M. *J. Org. Chem.* **1960**, *25*, 324.

The product is formed by an ene reaction between benzyne and the diene.

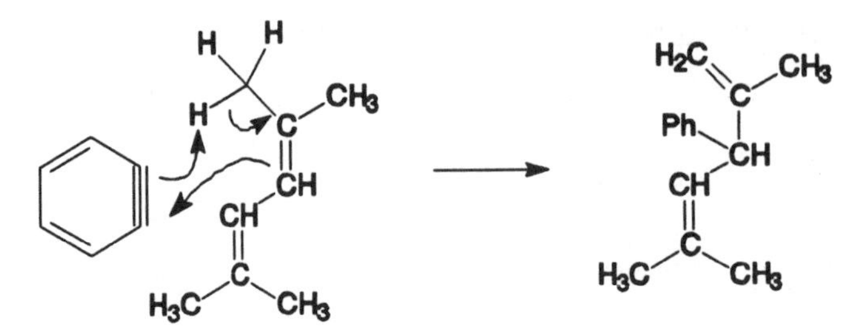

16. Copley, S. D.; Knowles, J. R. *J. Am. Chem. Soc.* **1985**, *107*, 5306. Also see Dagani, R. *Chem. Eng. News* **1984** (May 28), 26.

a. The rearrangement is a [3,3] sigmatropic rearrangement:

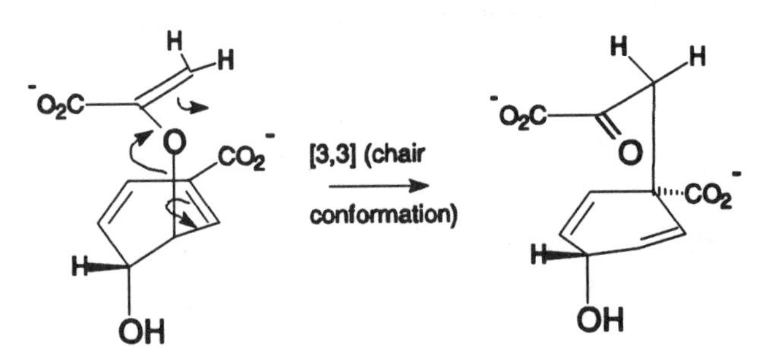

b. The diastereomers are shown here, where "T" represents ^3H.

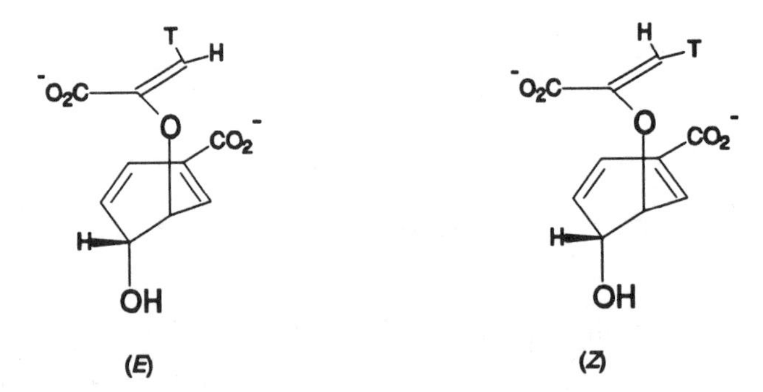

c.

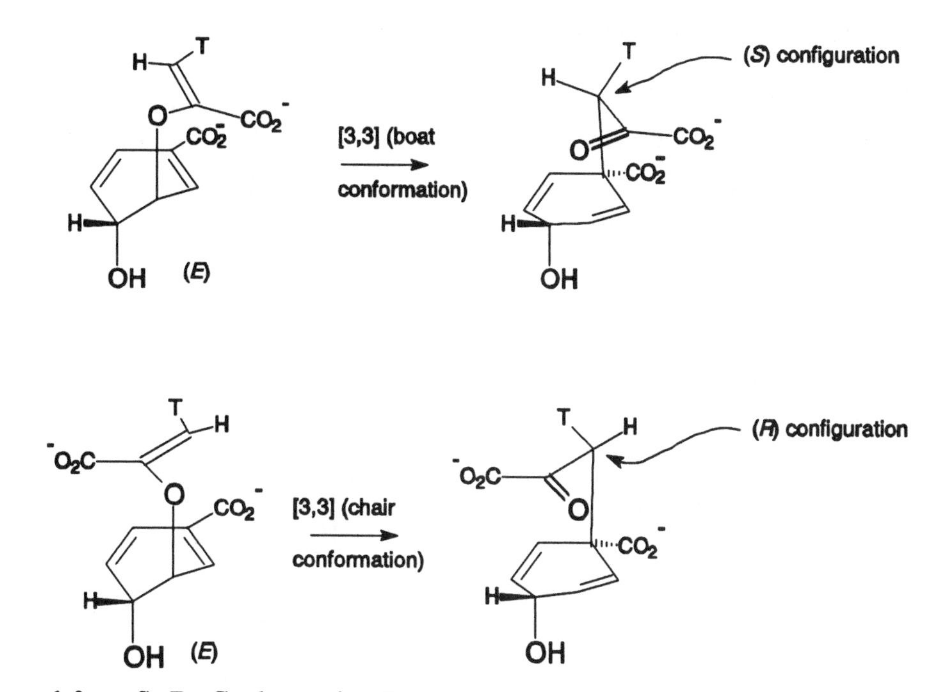

(Adapted from S. D. Copley and J. R. Knowles, *J. Am. Chem. Soc.* **107** (1985) 5306.)

d.

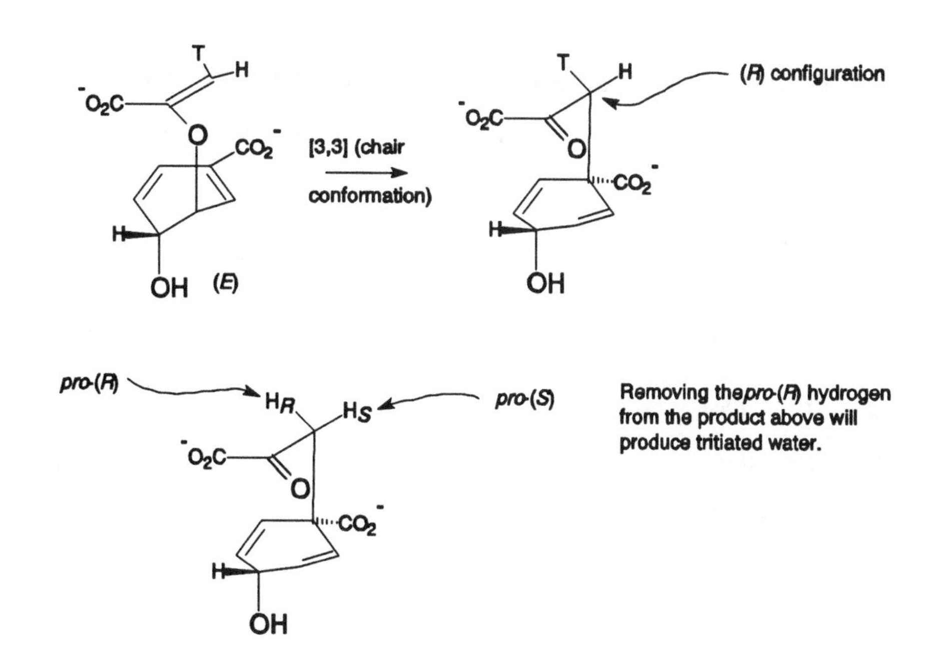

17. Conroy, H.; Firestone, R. A. *J. Am. Chem. Soc.* **1956**, *78*, 2290.

Products having molecular formula $C_{15}H_{16}O_4$ are consistent with (racemic) Diels-Alder adducts resulting from reaction of maleic anhydride and 6-allyl-2,6-dimethyl-2,4-cyclohexadieneone. Their formation suggests that the cyclohexadieneone is an intermediate in the thermal conversion of allyl 2,6-dimethylphenyl ether to 4-allyl-2,6-dimethylphenol by consecutive [3,3] sigmatropic rearrangements.

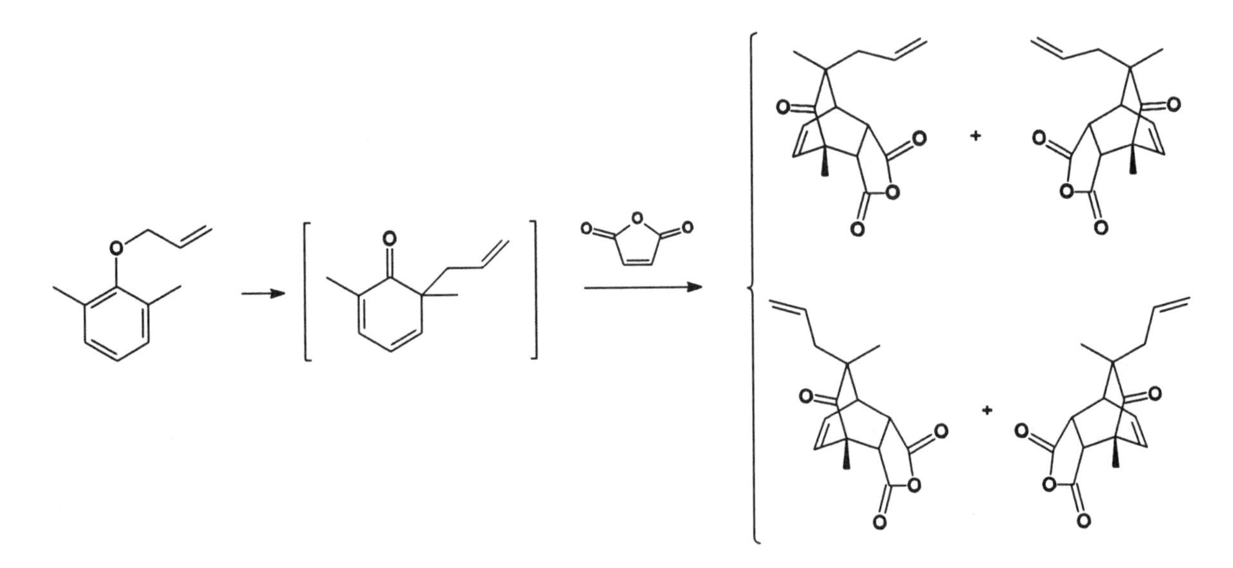

18. Heimgartner, H.; Hansen, H.-J.; Schmid, H. *Helv. Chim. Acta* **1972**, *55*, 1385.

The first is a disrotatory electrocyclic closure, followed by a [1,5] hydrogen shift.

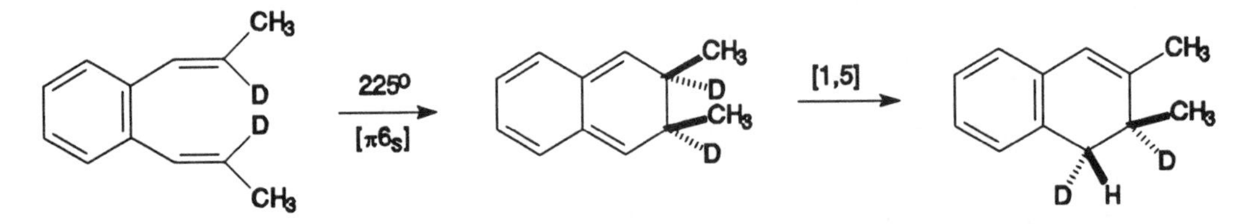

19. Greenwood, F. L. *J. Org. Chem.* **1959**, *24*, 1735. See also the discussion in DePuy, C. H.; King, R. W. *Chem. Rev.* **1960**, *60*, 431 (see p. 443).

The reactants equilibrate by a concerted transition reaction before elimination of acetic acid occurs.

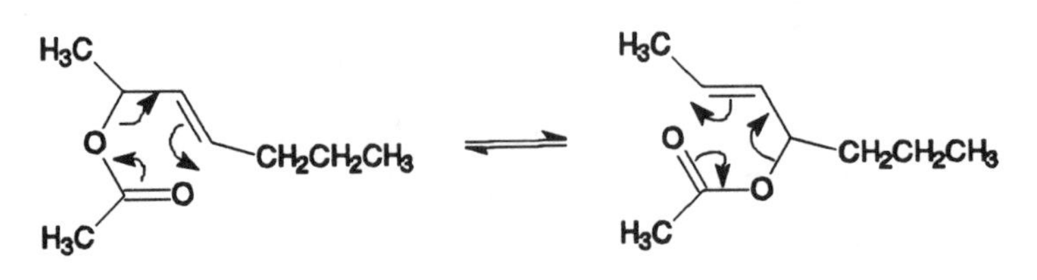

20. Agosta, W. C. *J. Am. Chem. Soc.* **1964**, *86*, 2638.

The product was formed by a Diels-Alder reaction (SS), so the configuration of the (-)-pentadienedioic acid must have been (*R*).

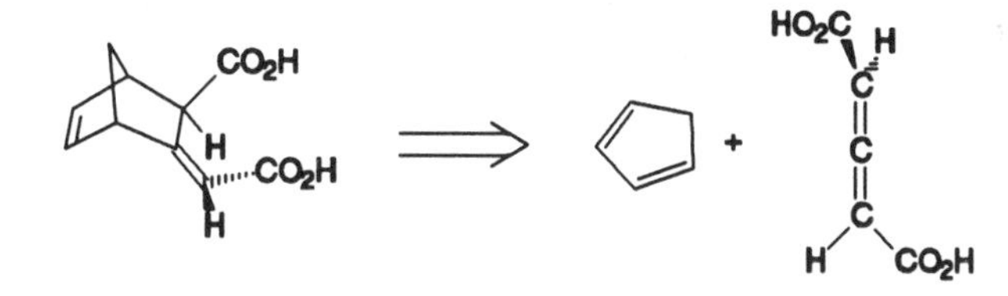

21. Baldwin, J. E.; Reddy, V. P.; Schaad, L. J.; Hess, Jr., B. A. *J. Am. Chem. Soc.* **1988**, *110*, 8555.

a. The rate constant for formation of **99** is $[1.1/(1.1 + 1.0) \times 8.17 \times 10^{-5}\ \text{sec}^{-1}] = 4.28 \times 10^{-5}\ \text{sec}^{-1}$. Similarly, the rate constant for formation of **100** is $[1.0/(1.1 + 1.0) \times 8.17 \times 10^{-5}\ \text{sec}^{-1}] = 3.89 \times 10^{-5}\ \text{sec}^{-1}$.

b. $k_H/k_D = 4.465/4.28 = 1.04$ for formation of **99**. For formation of **100**, $k_H/k_D = 4.465/3.89 = 1.15$.

c. These are α 2° kinetic isotope effects. The different values of k_H/k_D indicate that rehybridization of the orbitals (formally from sp^3 to sp^2) is not occurring to the same extent for the "inside" and "outside" hydrogens in the transition structure. This result is inconsistent with the usual model for an electrocyclic reaction in which both of the substituents on the terminal carbon atoms have rotated to the same extent in the transition structure. The authors of the paper cited below calculated that the hybridization of the orbital to the inner hydrogen atom is 74% *p* character in the transition structure, while the orbital to the outer hydrogen atom is 68% *p* character.

Chapter 12

1. Absorption onset of 375 nm = 76.3 kcal/mol; Fluorescence onset of 25,900 cm^{-1} $\times$ 0.00286 = 74.1 kcal/mol; phosphorescence onset 6,800 Å = 680 nm = 42 kcal/mol.

2. a. 5 x 10^{-5} moles of product/10^{-3} Einsteins = 0.05 for Φ_{prod}

   b. 1 - (0.3 + 0.5 + 0.05) = 0.15

3. Compound **O** is anthracene; compound **Q** is DDT. Data are unpublished results obtained by the author. Plotting ln I versus t in each case gives a k value of 2 x 10^3 sec^{-1} for [Q] = 0; k' values are 4.6 x 10^3 sec^{-1} for [Q] = 0.1 M and 5.8 x 10^3 sec^{-1} for [Q] = 0.16 M. Then plotting k'/k (equivalently, τ/τ') versus [Q] produces a Stern-Volmer plot with slope 11.81. Using a value of τ of $1/k$ gives a k_q value of 2.4 x 10^4 L mol^{-1} sec^{-1}.

4. Babu, M. K.; Rajasekaran, K.; Kannan, N.; Gnanasekaran, C. *J. Chem. Soc., Perkin Trans. 2* **1986**, 1721.

   pK_a^* - pK_a = 8.81, so pK_a^* = 2.72.

5. Pacifici, J. G.; Hyatt, J. A. *Mol. Photochem.* **1971**, *3*, 271.

   This is the ester analogue of a Norrish Type II reaction involving hydrogen abstraction, followed either by reverse hydrogen atom transfer after bond rotation to form the enantiomer or by fragmentation to the alkene and acid.

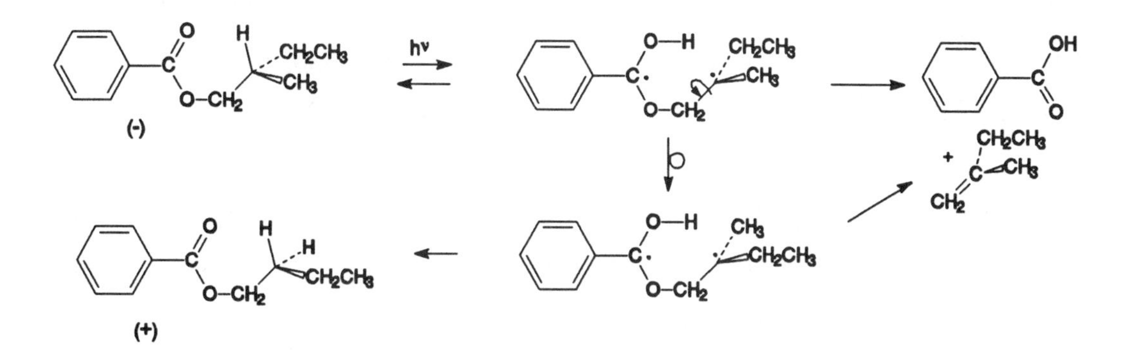

6. Frey, H. M.; Lister, D. H. *Mol. Photochem.* **1972**, *3*, 323.

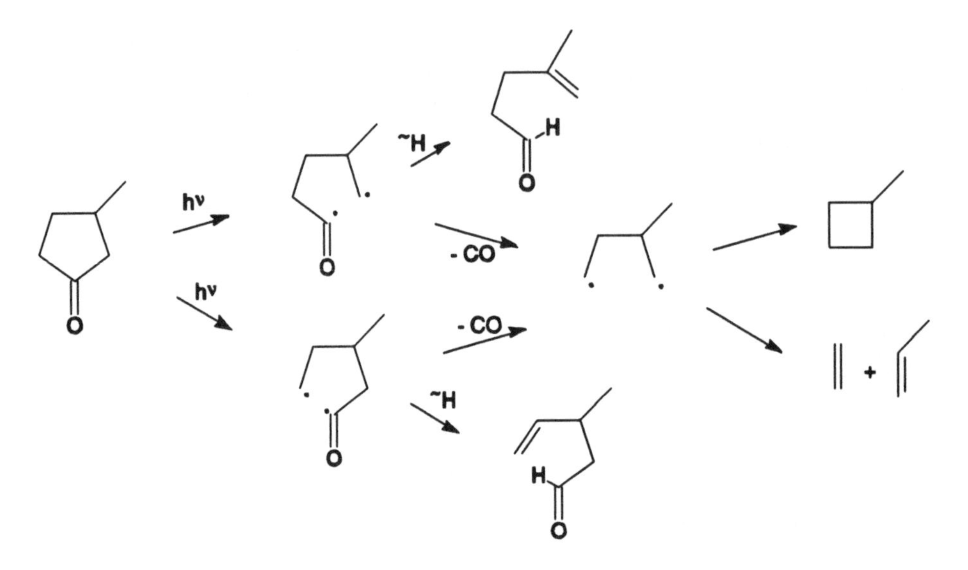

7. Arnold, B.; Donald, L.; Jurgens, A.; Pincock, J. A. *Can. J. Chem.* **1985**, *63*, 3140 and references therein.

[Q] (M)	Percent Triplets Quenched	Percent Singlets Quenched
1	0%	17%
.1	1%	67%
.01	5%	95%
.001	36%	100%

8. Tsuneishi, H.; Inoue, Y.; Hakushi, T.; Tai, A. *J. Chem. Soc. Perkin Trans. 2* **1993**, 457.
The ratio of isomers present at the photostationary state is $([E]/[Z])_{pss} = 0.23/0.77 = 0.30$.
Equation 12.52 is rewritten according to the symbolism used in this problem as follows:

$$([E]/[Z])_{pss} = (k_{dE}/k_{dZ}) \times (\epsilon_Z/\epsilon_E)$$

Rearranging terms and solving for (k_{dE}/k_{dZ}) yields a value of 0.98. In other words, the excited singlet state formed from either isomer has a nearly equal probability of relaxing to the more stable (*Z*) diasteromer or to the less stable (*E*) diasteromer.

9. Alumbaugh, R. L.; Pritchard, G. O.; Rickborn, B. *J. Phys. Chem.* **1965**, *69*, 3225.
α-Cleavage leads to a diradical that can close to the isomer of the reactant, can undergo intramolecular hydrogen atom transfer to form isomeric aldehydes, or can lose CO to form another diradical that can close to isomeric 1,2-dimethylcyclopentanes or undergo intramolecular hydrogen abstraction to form isomeric alkenes.

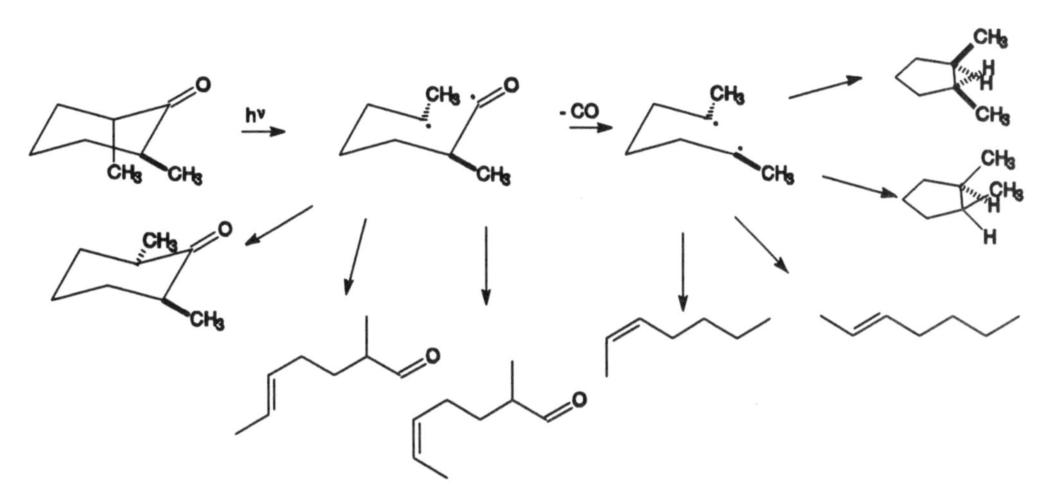

10. Ohloff, G.; Klein, E.; Schenck, G. O. *Angew. Chem.* **1961**, *73*, 578. See also Schönberg, A. *Preparative Organic Photochemistry*; Springer-Verlag: New York, 1968; p. 377.

The other product is 2,6-dimethyl-1-octene-3,8-diol. It is formed by the same mechanism but with the other orientation.

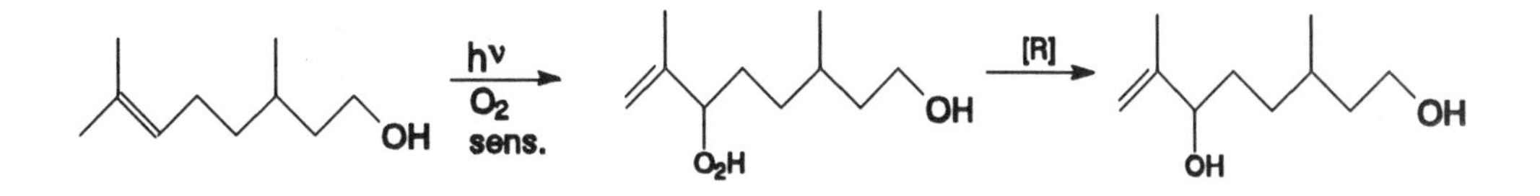

11. a. Kropp, P. J. *Mol. Photochem.* **1978-79**, *9*, 39 and references therein.

In each case, the D atom is on the carbon adjacent to the carbon bearing the double bond or the methoxy group.

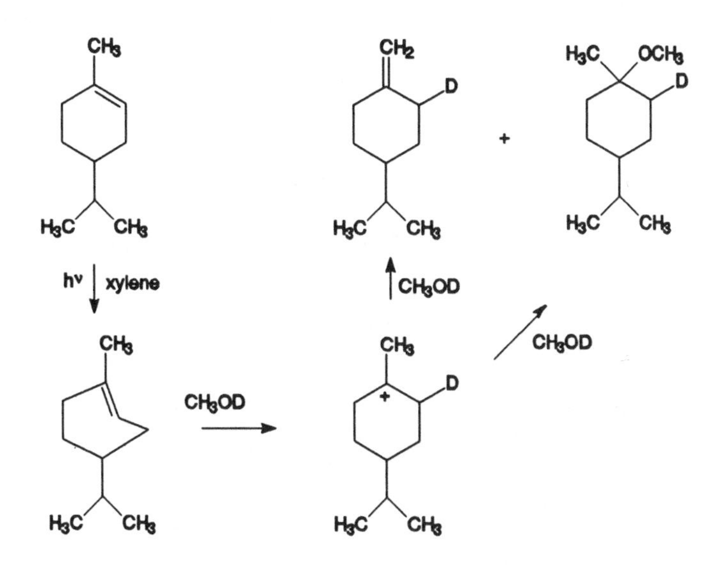

b. Yang, N. C.; Jorgenson, M. J. *Tetrahedron Lett.* **1964**, 1203.

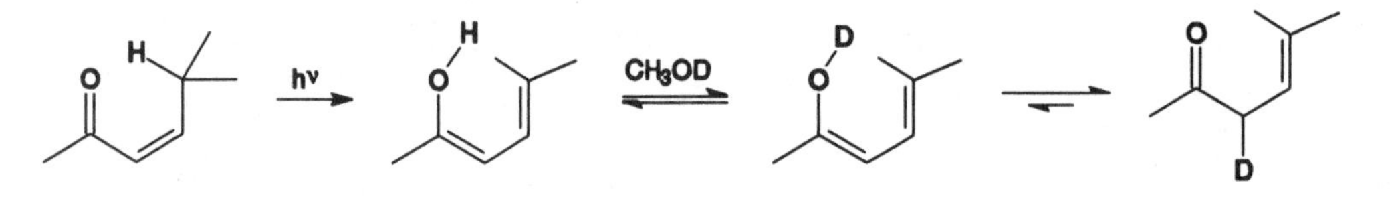

The deuterium label is on C3 because tautomerization of the dienol-OD that is formed by equilibration of the dienol-OH with solvent.

12. Carless, H. A. J. *J. Chem. Soc., Perkin Trans. 2* **1974**, 834.

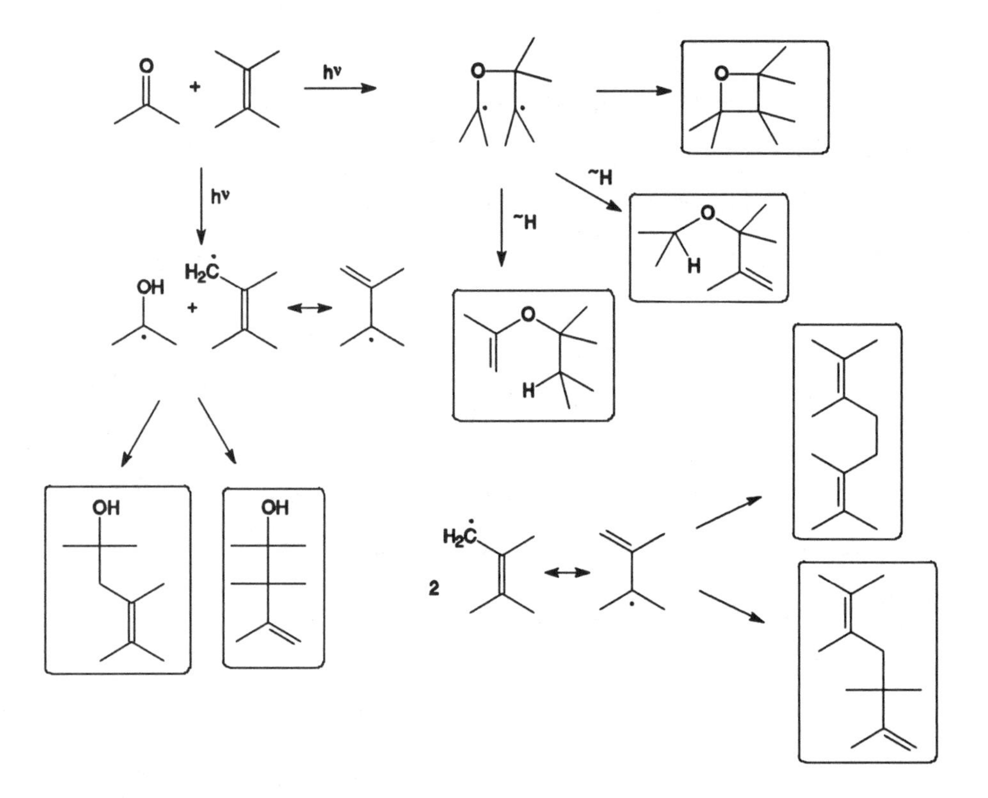

13. Zimmerman, H. E.; Sandel, V. R. *J. Am. Chem. Soc.* **1963**, *85*, 915. Also see DeCosta, D. P.; Pincock, J. A. *J. Am. Chem. Soc.* **1993**, *115*, 2180.

The first reaction occurs predominantly by a radical pathway, while the second is predominantly a photosolvolysis. However, there is evidence for some radical reaction in the photochemistry of the meta isomer also. (Yields shown are the average values obtained in two experiments.)

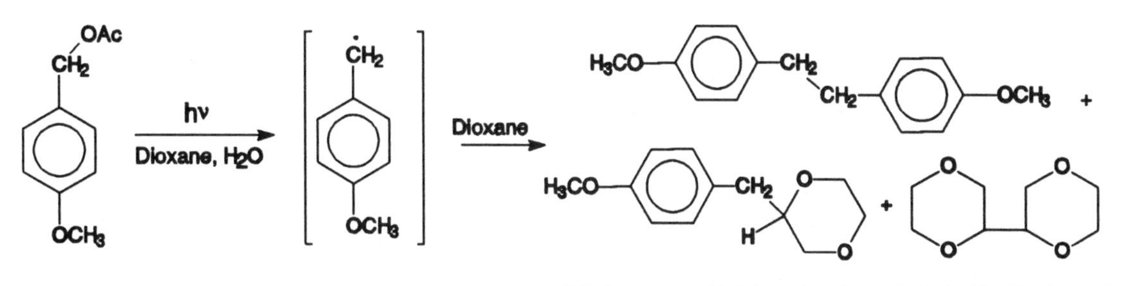

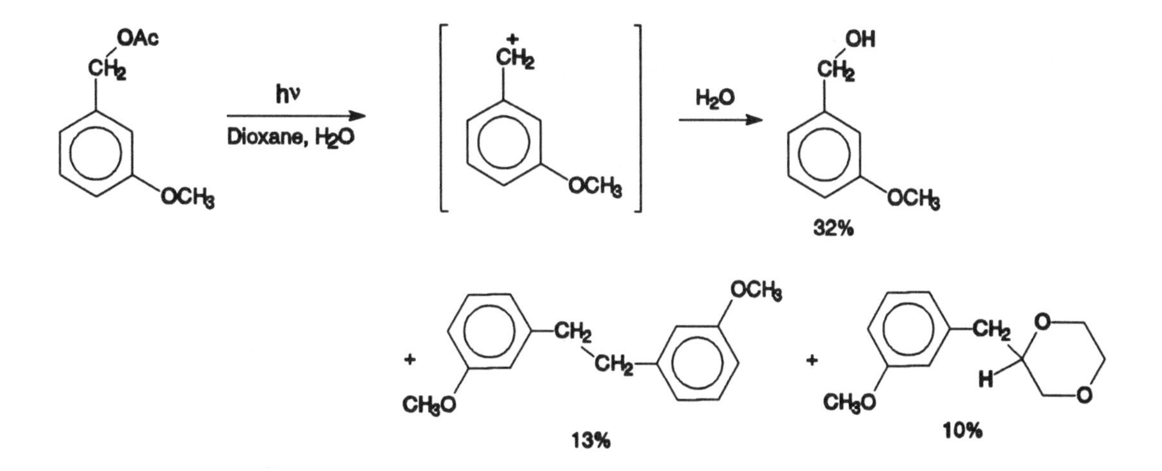

14. a. Butenandt, A.; Poschmann, L. *Chem. Ber.* **1944**, *77B*, 394.

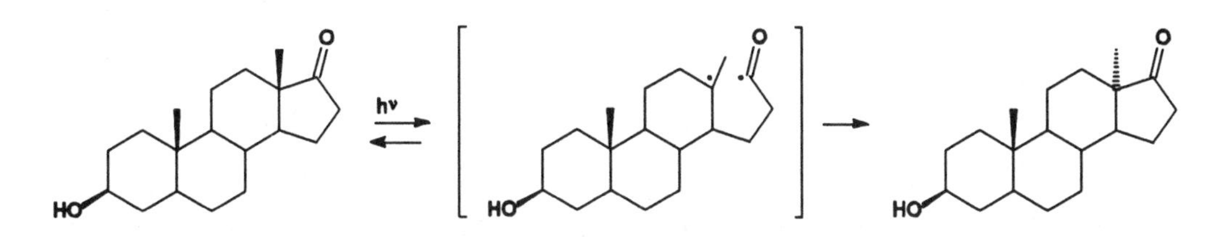

b. Quinkert, G. *Angew. Chem., Int. Ed. Engl.* **1962**, *1*, 166.

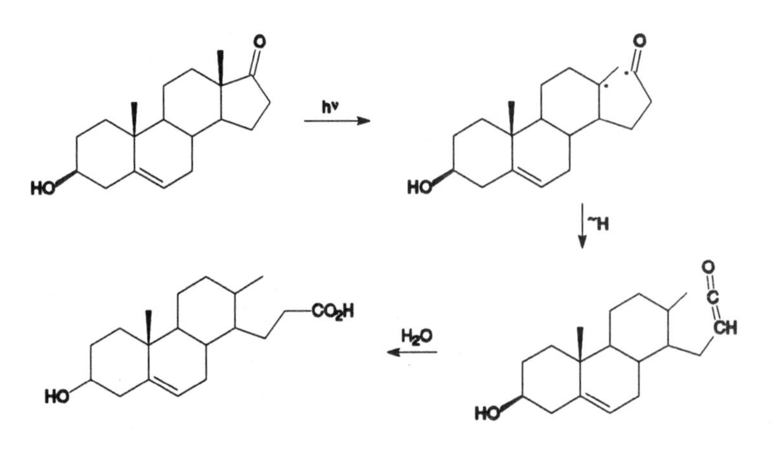

c. Quinkert, G. *Angew. Chem., Int. Ed. Engl.* **1962**, *1*, 166.

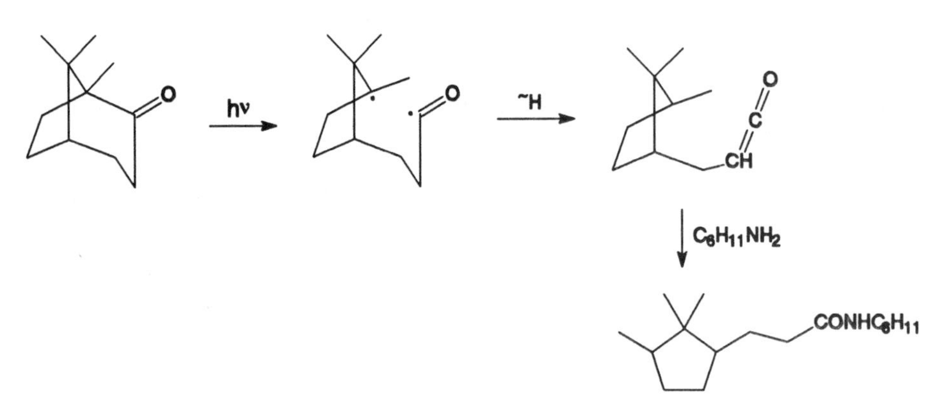

d. Medary, R. T.; Gano, J. E.; Griffin, C. E. *Mol. Photochem.* **1974**, *6*, 107.

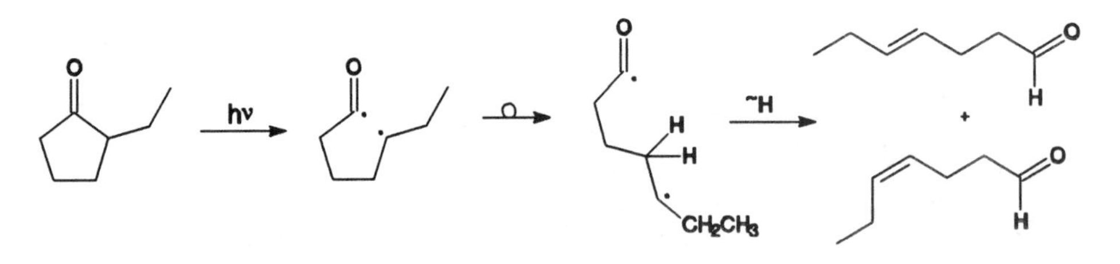

15. a. Barnard, M.; Yang, N. C. *Proc. Chem. Soc.* **1958**, 302.

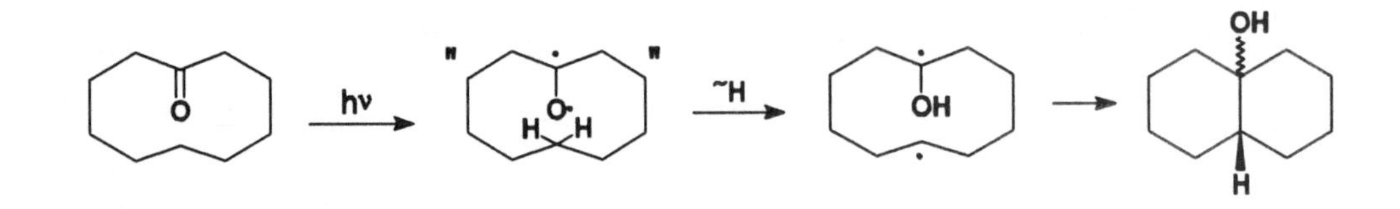

b. Wolff, S.; Schreiber, W. L.; Smith, III, A. B.; Agosta, W. C. *J. Am. Chem. Soc.* **1972**, *94*, 7797.

The reaction occurs by abstraction of a hydrogen atom from the methoxy group by the ß-carbon atom of the enone.

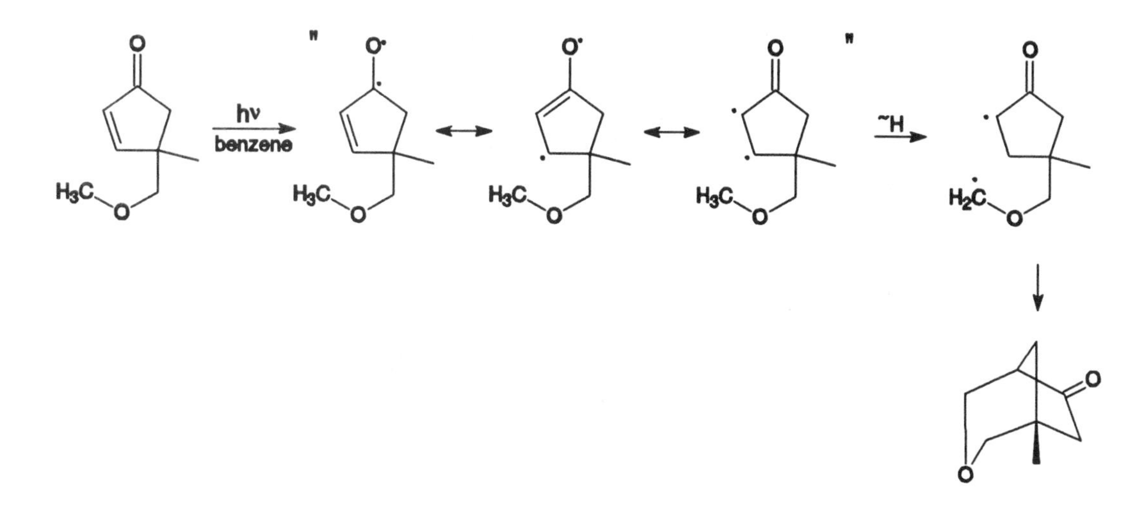

16. a. Agosta, W. C.; Smith, III, A. B. *J. Am. Chem. Soc.* **1971**, *93*, 5513. See also Wolff, S.; Schreiber, W. L.; Smith, III, A. B.; Agosta, W. C. *J. Am. Chem. Soc.* **1972**, *94*, 7797. Other products are also formed in the reaction.

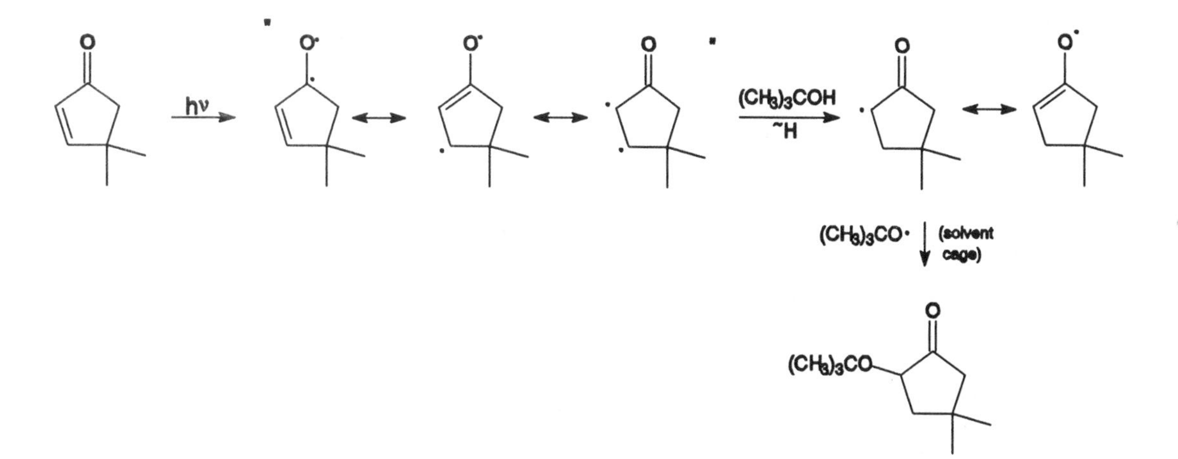

This reaction apparently occurs through hydrogen atom abstraction by the ß-carbon atom of the enone. Photoreduction would be expected to result from irradiation of the reactant in a 1° or 2° alcohol.

b. Göth, H.; Cerutti, P.; Schmid, H. *Helv. Chem. Acta* **1965**, *48*, 1395.

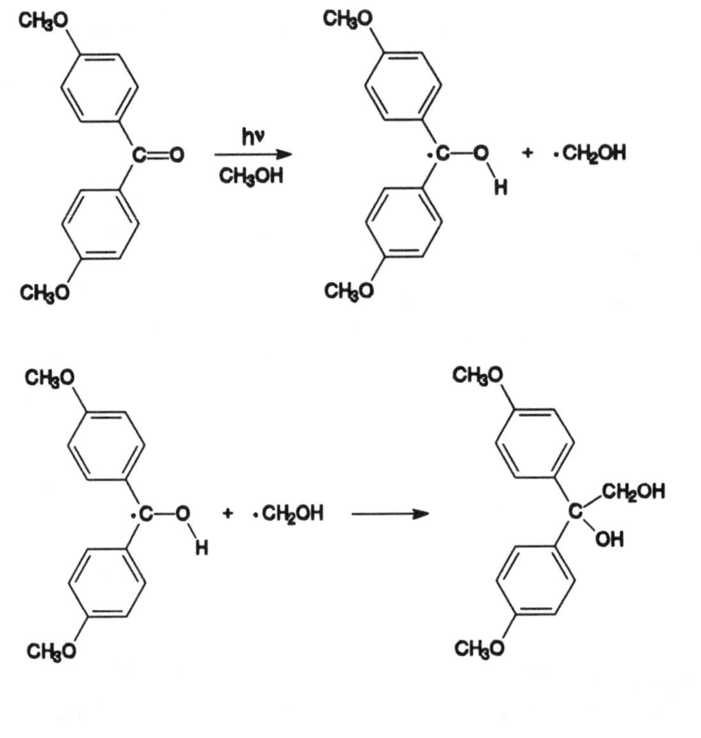

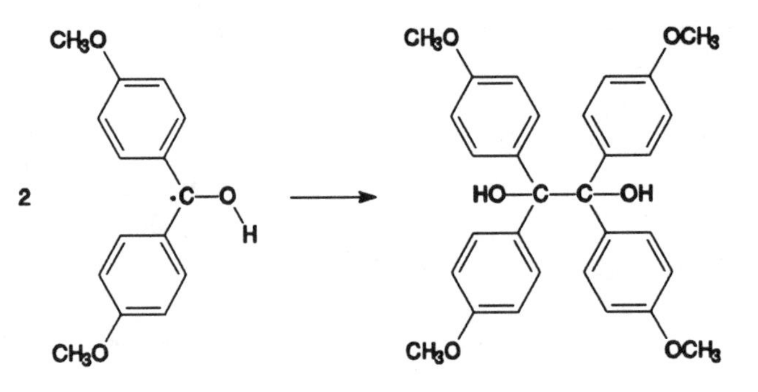

17. a. Anderson, J. C.; Reese, C. B. *J. Chem. Soc.* **1963**, 1781.

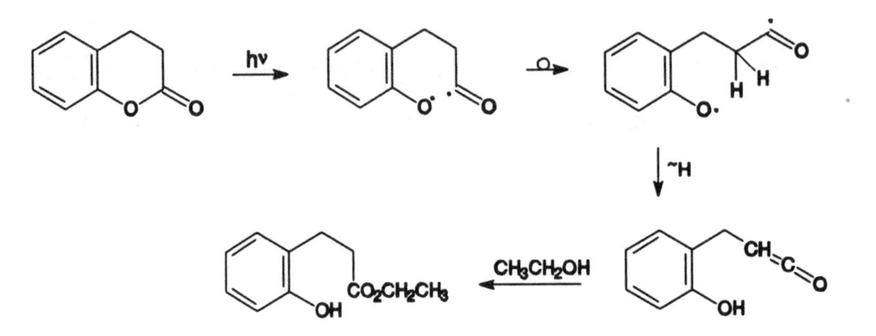

b. Anderson, J. C.; Reese, C. B. *J. Chem. Soc.* **1963**, 1781.

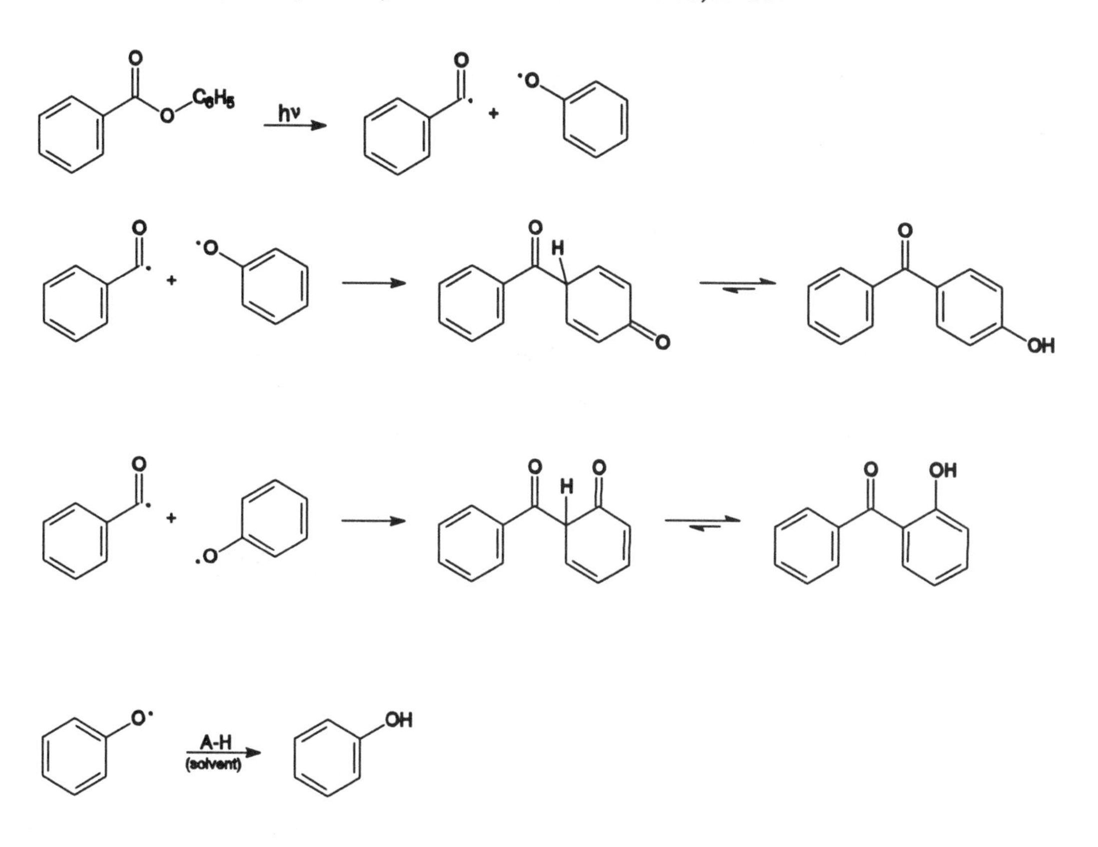

18. Padwa, A.; Alexander, E.; Niemcyzk, M. *J. Am. Chem. Soc.* **1969**, *91*, 456. See also Padwa, A. *Acc. Chem. Res.* **1971**, *4*, 48.

In isopropyl alcohol solution, intermolecular hydrogen abstraction leads to radicals that combine to give the product observed.

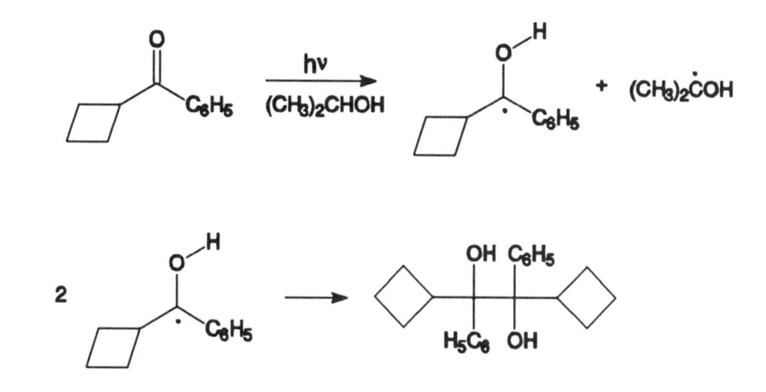

In benzene solution, abstraction of hydrogen from solvent is difficult, so intramolecular hydrogen abstraction occurs instead.

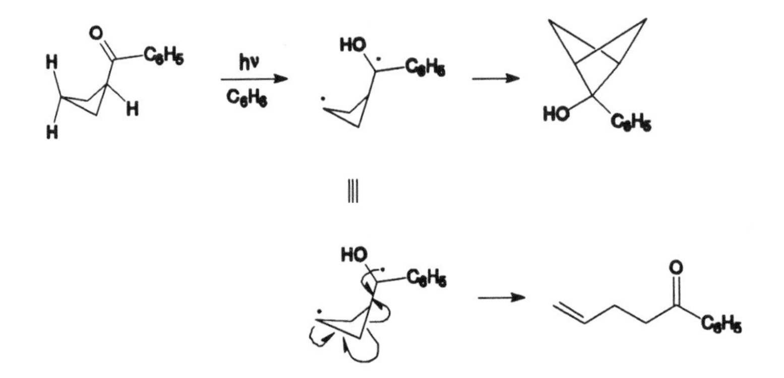

19. a. Bahurel, Y.; Descotes, G.; Pautet, F. *Compt. Rend.* **1970**, *270*, 1528; Bahurel, Y.; Pautet, F.; Descotes, G. *Bull. Soc. Chim. France*, **1971**, *6*, 2222.

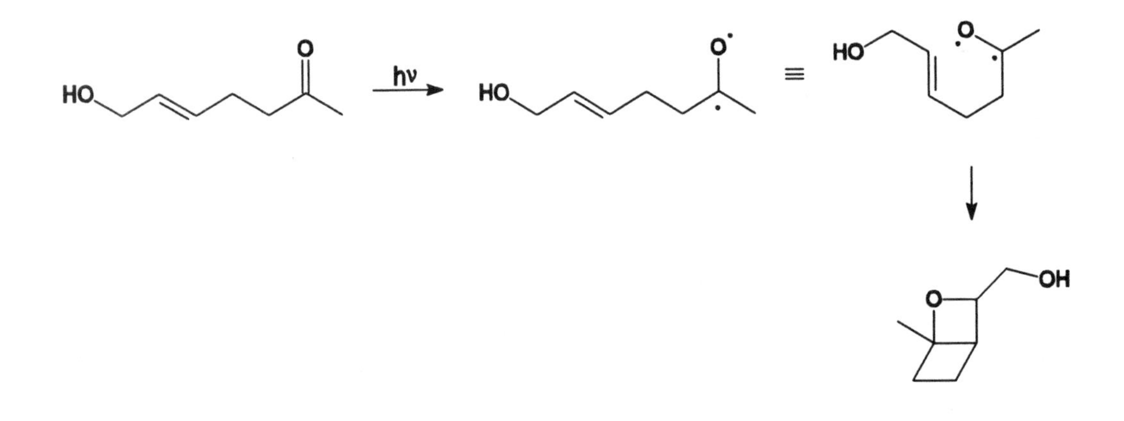

b. Nobs, F.; Burger, U.; Schaffner, K. *Helv. Chim. Acta* **1977**, *60*, 1607.

Hydrogen abstraction form the methyl group of the propenyl moiety by the α-carbon atom of the enone group leads to a diradical that closes to the product.

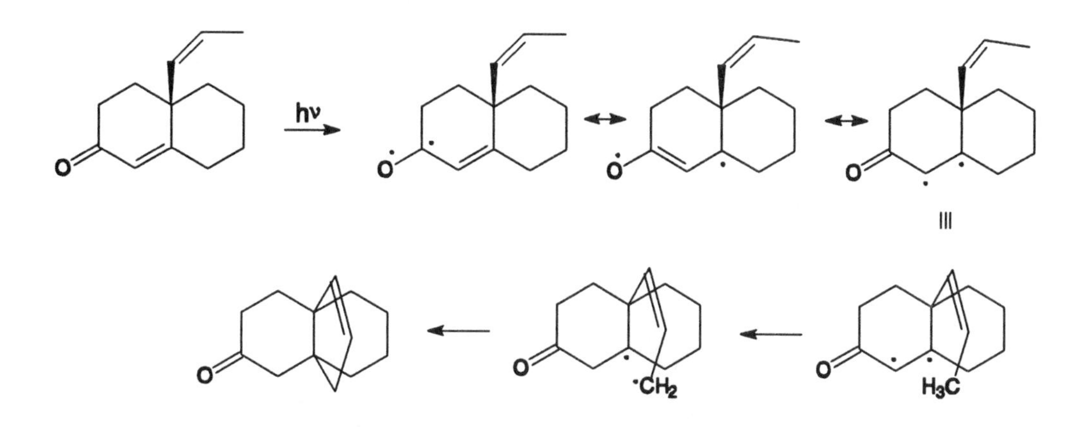

20. Turro, N. J.; Weiss, D. S. *J. Am. Chem. Soc.* **1968**, *90*, 2185. See also Dawes, K.; Dalton, J. C.; Turro, N. J. *Mol. Photochem.* **1971**, *3*, 71.

There is a stereochemical requirement for Norrish Type II abstraction. Specifically, only a γ-hydrogen atom in the plane of the carbonyl group (that is, near the half-empty nonbonding orbital of the n,π* excited ketone) can be abstracted. In the absence of a γ-hydrogen atom in this position, only Norrish Type I cleavage occurs, leading to photoepimerization.

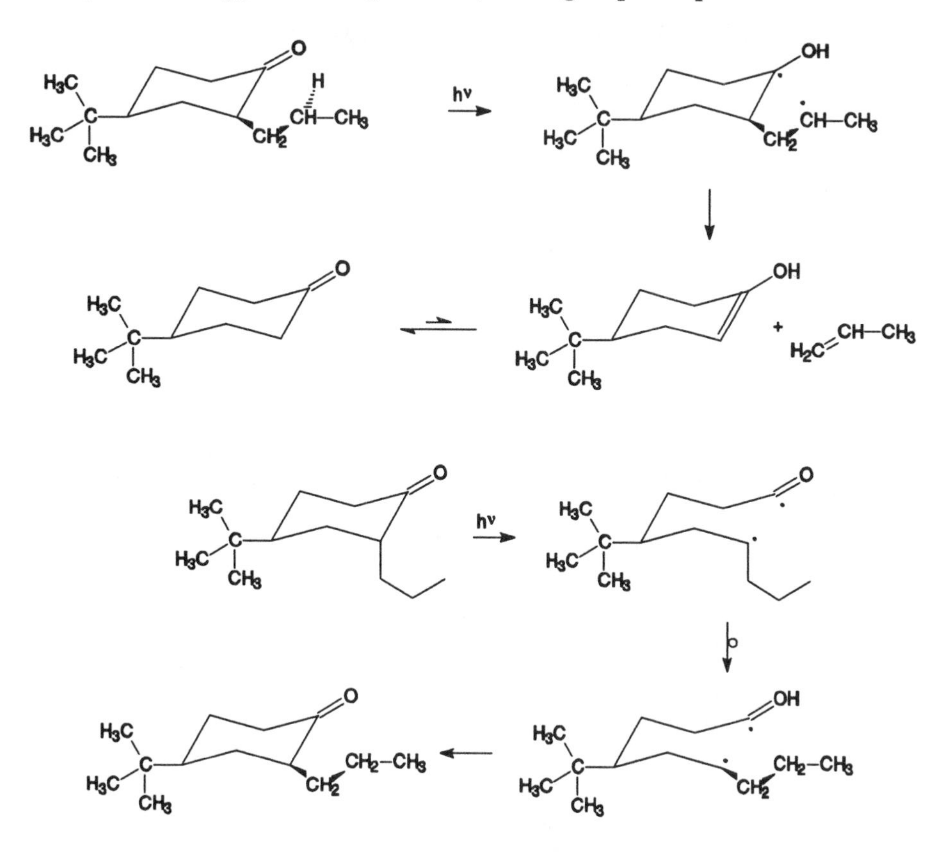

21. a. Padwa, A.; Eisenberg, W. *J. Am. Chem. Soc.* **1970**, *92*, 2590.

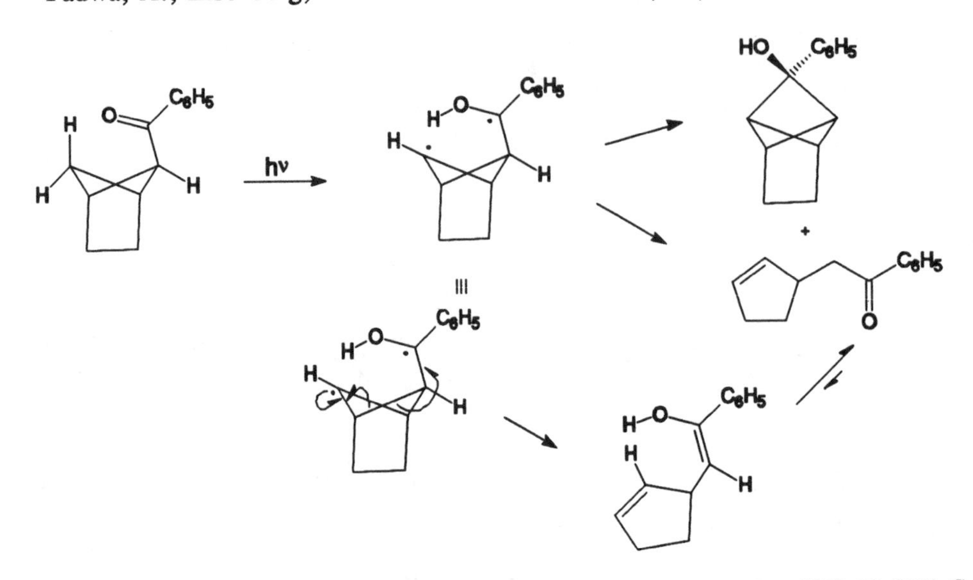

b. Gagosian, R. B.; Dalton, J. C.; Turro, N. J. *J. Am. Chem. Soc.* **1970**, *92*, 4752; see also Turro, N. J.; Dalton, J. C.; Dawes, K.; Farrington, G.; Hautala, R.; Morton, D.; Niemczyk, M.; Schore, N. *Acc. Chem. Res.* **1972**, *5*, 92.

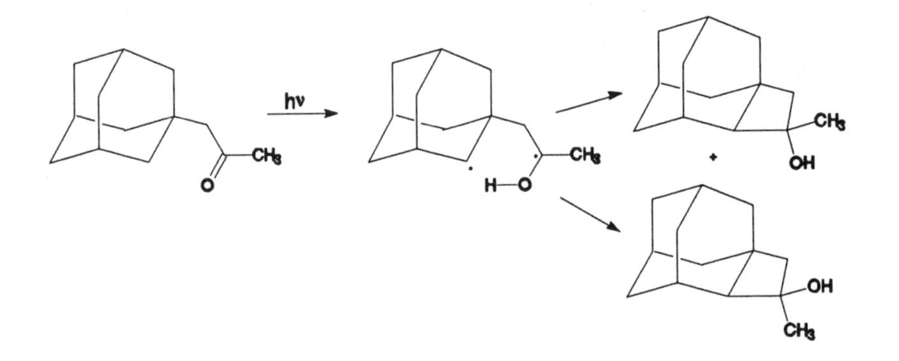

c. Padwa, A.; Alexander, E.; Niemcyzk, M. *J. Am. Chem. Soc.* **1969**, *91*, 456; see also Padwa, A. *Acc. Chem. Res.* **1971**, *4*, 48.

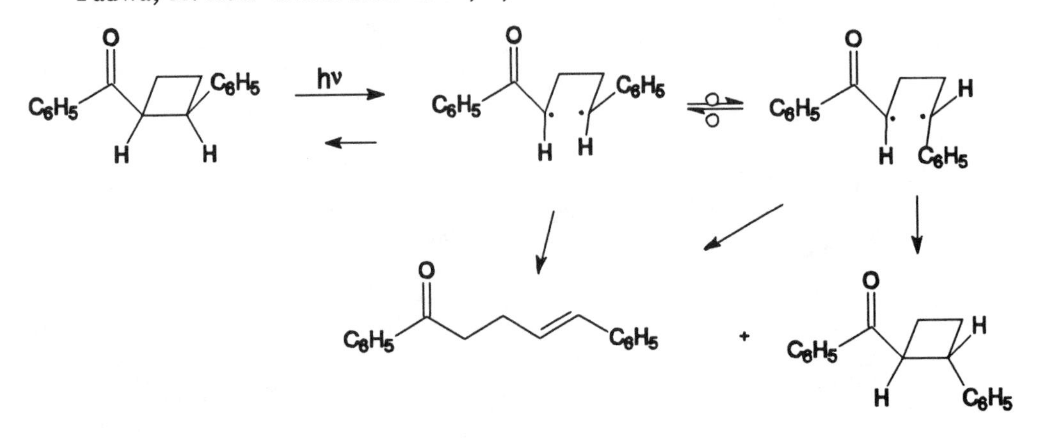

d. Arnold, B. J.; Mellows, S. M.; Sammes, P. G.; Wallace, T. W. *J. Chem. Soc., Perkin Trans. 1* **1974**, 401. See also Durst, T.; Kozma, E. C.; Charlton, J. L. *J. Org. Chem.* **1985**, *50*, 4829.

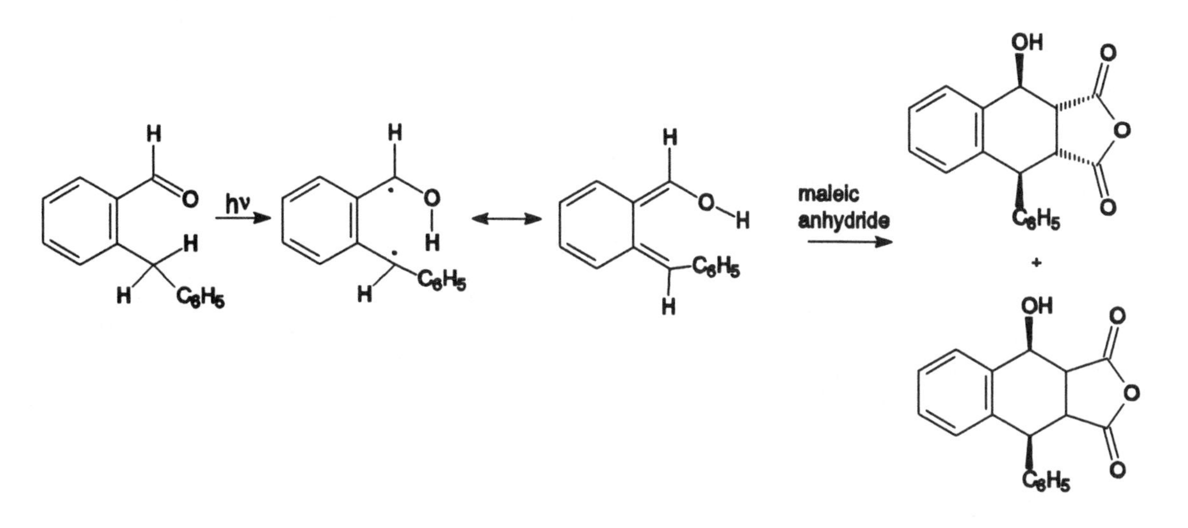

e. Wolff, S.; Schreiber, W. L.; Smith, III, A. B.; Agosta, W. C. *J. Am. Chem. Soc.* **1972**, *94*, 7797.

Abstraction of the 3° hydrogen from the 3-methylbutyl substituent by the β carbon atom of the enone leads to a biradical. Disproportionation by hydrogen abstraction from a carbon adjacent to the radical center on the alkyl chain leads to the first two products; closure of the diradical leads to the third.

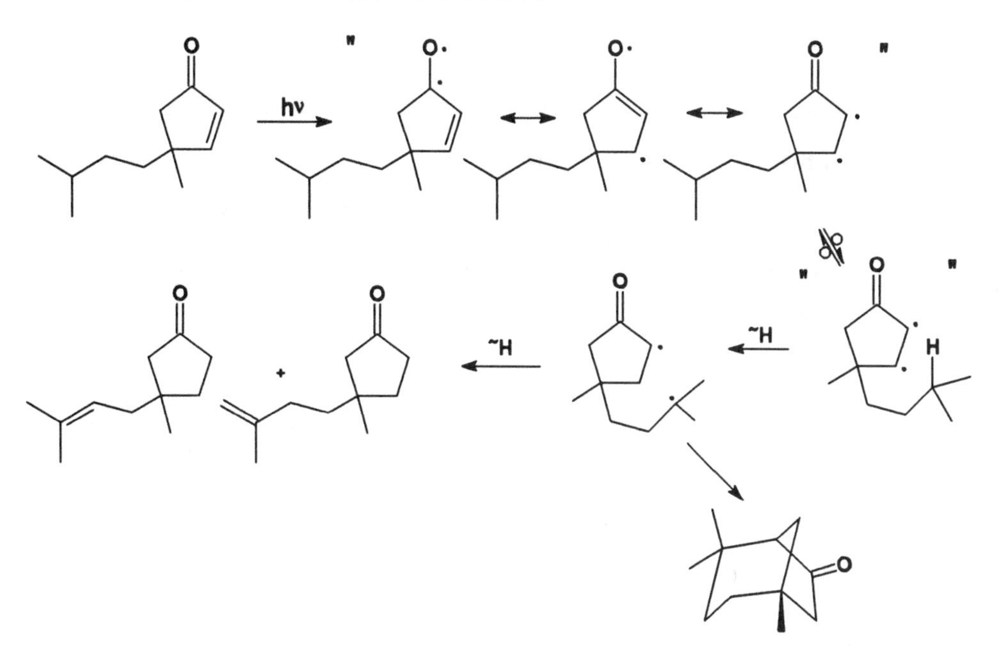

22. Gilbert, A.; Taylor, G. N.; bin Samsudin, M. W. *J. Chem. Soc., Perkin Trans. 1* **1980**, 869.
The intermediate is a 1,2-photoadduct. Although not discussed explicitly in the literature reference cited, a possible mechanism for the conversion of the adduct to the final product involves acid-catalyzed loss of methanol, rearrangement (ring opening) of the carbocation, and proton loss.

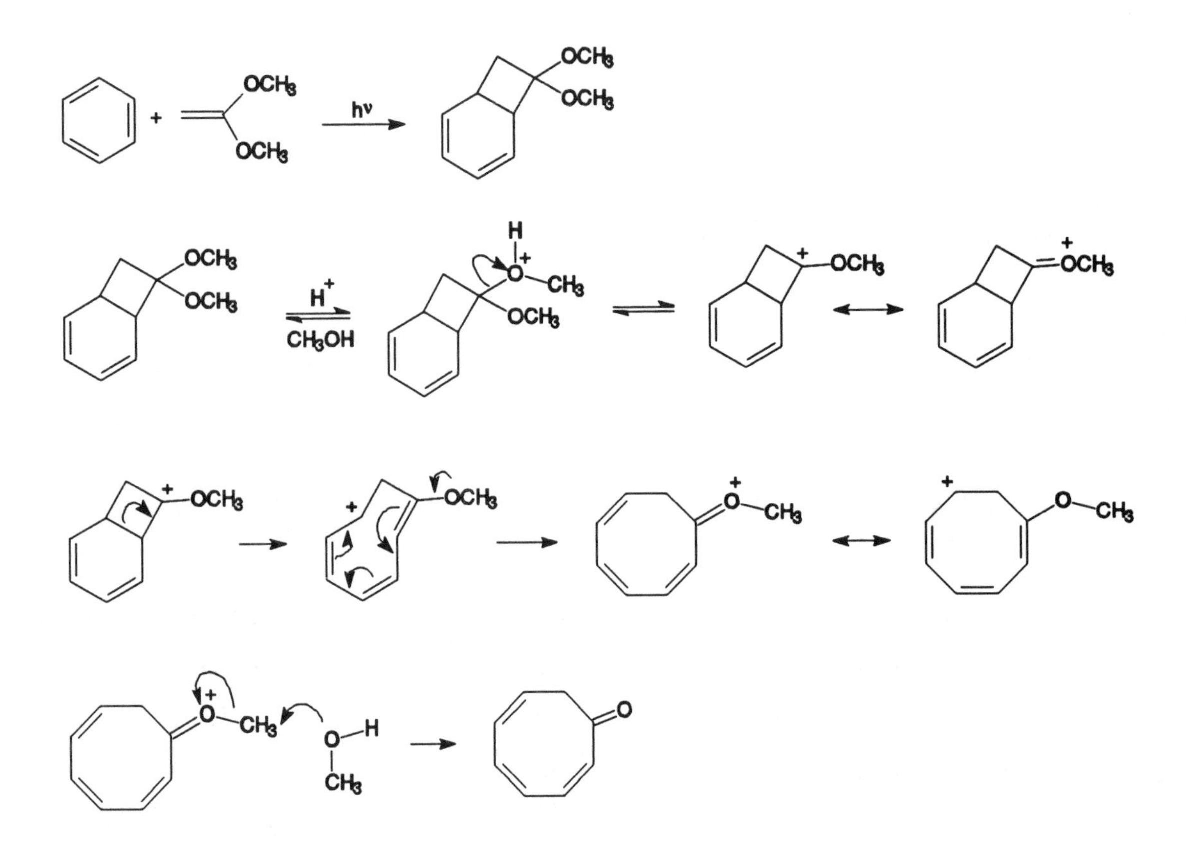

23. Morrison, H.; Pajak, J.; Peiffer, R. *J. Am. Chem. Soc.* **1971**, *93*, 3978.
The data suggest that the benzene ring absorbs the excitation, undergoes intersystem crossing to the aromatic triplet, and transfers energy to the olefin (intramolecular triplet sensitization). The olefinic triplet then decays to the cis and trans isomers of the starting material.